Application Design for
Wearable Computing

Synthesis Lectures on Mobile and Pervasive Computing

Editor

Mahadev Satyanarayanan, *Carnegie Mellon University*

Mobile computing and pervasive computing represent major evolutionary steps in distributed systems, a line of research and development that dates back to the mid-1970s. Although many basic principles of distributed system design continue to apply, four key constraints of mobility have forced the development of specialized techniques. These include unpredictable variation in network quality, lowered trust and robustness of mobile elements, limitations on local resources imposed by weight and size constraints, and concern for battery power consumption. Beyond mobile computing lies pervasive (or ubiquitous) computing, whose essence is the creation of environments saturated with computing and communication yet gracefully integrated with human users. A rich collection of topics lies at the intersections of mobile and pervasive computing with many other areas of computer science.

RFID Explained
Roy Want
2006

Controlling Energy Demand in Mobile Computing Systems
Carla Schlatter Ellis
2007

Application Design for Wearable Computing
Dan Siewiorek, Asim Smailagic, and Thad Starner
2008

Application Design for Wearable Computing
Dan Siewiorek
Asim Smailagic
Thad Starner

ISBN: 978-3-031-01348-5   paperback
ISBN: 978-3-031-02476-4   ebook

DOI: 10.1007/978-3-031-02476-4

A Publication in the Springer series

SYNTHESIS LECTURES ON MOBILE AND PERVASIVE COMPUTING #3

Lecture #3

Series Editor: Mahadev Satyanarayanan, Carnegie Mellon University

Series ISSN

ISSN 1933-9011   print
ISSN 1933-902X   electronic

Application Design for Wearable Computing

Dan Siewiorek and Asim Smailagic
Carnegie Mellon University

Thad Starner
Georgia Institute of Technology

SYNTHESIS LECTURES ON MOBILE AND PERVASIVE COMPUTING #3

ABSTRACT

The confluence of decades of computer science and computer engineering research in multimodal interaction (e.g., speech and gesture recognition), machine learning (e.g., classification and feature extraction), software (e.g., web browsers, distributed agents), electronics (e.g., energy-efficient microprocessors, head-mounted displays), design methodology in user-centered design, and rapid prototyping have enabled a new class of computers—wearable computers. The lecture takes the viewpoint of a potential designer or researcher in wearable computing. Designing wearable computers requires attention to many different factors because of the computer's closeness to the body and its use while performing other tasks. For the purposes of discussion, we have created the UCAMP framework, which consists of the following factors: user, corporal, attention, manipulation, and perception. Each of these factors and their importance is described. A number of example prototypes developed by the authors, as well as by other researchers, are used to illustrate these concepts. Wearable computers have established their first foothold in several application domains, such as vehicle and aircraft maintenance and manufacturing, inspection, language translation, and other areas.

The lecture continues by describing the next step in the evolution of wearable computers, namely, context awareness. Context-aware computing takes into account a user's state and surroundings, and the mobile computer modifies its behavior based on this information. A user's context can be quite rich, consisting of attributes such as physical location, physiological state, personal history, daily behavioral patterns, and so forth. If a human assistant were given such context, he or she would make decisions in a proactive fashion, anticipating user needs, and acting as a proactive assistant. The goal is to enable mobile computers to play an analogous role, exploiting context information to significantly reduce demands on human attention. Context-aware intelligent agents can deliver relevant information when a user needs that information. These data make possible many exciting new applications, such as augmented reality, context-aware collaboration, and augmented manufacturing.

The combined studies and research reported in this lecture suggest a number of useful guidelines for designing wearable computing devices. Also included with the guidelines is a list of questions that designers should consider when beginning to design a wearable computer. The research directions section emphasizes remaining challenges and trends in the areas of user interface, modalities of interaction, and wearable cognitive augmentation. Finally, we summarize the most important challenges and conclude with a projection of future directions in wearable computing.

KEYWORDS

wearable computers, wearability design, user interface, manipulation devices, interaction modalities, context-aware computing, cognitive augmentation

Contents

CHAPTER 1

Introduction

Mobile access is the gating technology required to make information available at any place and at any time. Its application domains range from inspection, maintenance, manufacturing, and navigation to on-the-move collaboration, position sensing, and real-time speech recognition and language translation. In the course of developing wearable systems to support these applications, we have identified or refined several conceptual frameworks regarding personal computing.

At the core of these ideas is the notion that wearable computers should seek to merge the user's information space with his or her workspace. Information tools such as wearable computers must blend seamlessly with existing work environments, providing as little distraction as possible. This requirement often leads researchers to investigate replacements for the traditional console interfaces such as a keyboard or mouse, which generally require a fixed physical relationship between the user and the device. Identifying effective interaction modalities for wearable computers and accurately modeling user tasks in the supporting software are among the most significant challenges faced by wearable system designers. Because wearable computers represent a new paradigm in computing, there is no consensus on the mechanical/software human–computer interface or the capabilities of these systems.

In offices, computers have become a primary tool, allowing workers to access the information they need to perform their jobs. Accessing information is more difficult for mobile users, however. With current computer interfaces, the user must focus both physically and mentally on the computing device instead of the environments. In a mobile environment, such interfaces may interfere with the user's primary task. Yet many mobile tasks could benefit from computer support. Distractions are even more of a problem when they occur in mobile environments than in desktop environments because the user is often preoccupied with walking, driving, or other real-world interactions. The focus of this lecture is the design of wearable computers that augment, instead of interfere with, the user's tasks.

Body-worn computers that provide hands-free operation offer compelling advantages in many applications. Wearable computers deal in information rather than programs, becoming tools in the user's environment—much like a pencil or a reference book. The wearable computer provides portable access to information. Furthermore, the information can be accumulated automatically by the system as the user interacts with and modifies the environment, thereby eliminating the costly and error-prone process of transferring information to a central database later. When combined with pervasive computing, wearable computers will provide access to the right information at the right place and at the right time.

Wearable computing brings the power of a pervasive computing environment to a person by placing computing and sensory resources on the user in an unobtrusive way. These computers can be specialized and modular, like items of clothing. Unlike laptops or handheld computers, wearable computers offer many new models to interact beyond keyboards and touch screens, in a natural, intuitive way, such as sound and tactile feedback. In addition, wearables can easily be reconfigured to meet specific needs of applications.

Every wearable computer system must be viewed from three different axes: the human, the computer, and the application. Within each of these axes, there are difficult problems that must be solved. The human axis emphasizes wearability, which is defined as the interaction between the human body and the wearable object. Dynamic wearability includes the human body in motion. Design for wearability considers the physical shape of objects and their active relationship with the human form. Researchers explored history and cultures, including topics such as clothing, costumes, protective wearables, and carried devices. These studies of physiology, biomechanics, and movement were codified into guidelines for designing wearable systems. User comfort is a critical design consideration in many applications. New technologies such as smart textiles will significantly improve the functionality and ergonomics of wearable computers. The computer axis deals with the problems related to construction of a system with particular size, power consumption, and user interface software. The application axis emphasizes mobile application design challenges and efficient mapping of problem-solving capabilities to application requirements. Wearable computers have established their first foothold in several application domains, such as vehicle and aircraft maintenance and manufacturing, inspection procedures, language translation, and other areas.

The key research groups in academia worldwide today are at Carnegie Mellon, Columbia University, Georgia Tech, and MIT in the United States; Bremen University, Darmstadt University, ETH Zurich, and Lancaster University in Europe; University of South Australia, and NARA in Japan. WearIT@Work is a large wearable computing consortium of companies and universities in Europe, focusing on four application areas: mobile maintenance, emergency rescue, health care, and production. The project, which has 36 partners, is funded by the European Union. One of the main goals is to deploy wearable computers in real applications, such as aircraft maintenance, car manufacturing, firefighters' assistance, and health care.

Commercialization of general-purpose wearable computers so far has had limited success. Commercial companies include Xybernaut, CDI, and VIA Inc. Among commercial products are Seiko's Rupiter wristwatch computer and Panasonic's wearable brick computer coupled with a handheld or arm-worn touch screen. More recently, Sony and OQO have been successful in selling their wearable PCs. Symbol's success story with their wrist-mounted computer is elaborated in this lecture. To commemorate the first decade of the International Symposium on Wearable Computing, a brochure was created providing a retrospective of wearable systems and early experimentation with this technology. The brochure can be viewed at http://www.ices.cmu.edu/asim/iswc.pdf.

The next step in the evolution of wearable computers is context awareness. Context-aware computing takes into account a user's state and surroundings, and the mobile computer modifies its behavior based on this information. A user's context can be quite rich, consisting of attributes such as physical location, physiological state (such as body temperature, heart rate, and skin resistance), emotional state (such as angry, distraught, or calm), personal history, daily behavioral patterns, and so forth. If a human assistant were given such context, he or she would make decisions in a proactive fashion, anticipating user needs. In making these decisions, the assistant would typically not disturb the user at inopportune moments except in an emergency. The goal is to enable mobile computers to play an analogous role, exploiting context information to significantly reduce demands on human attention. Context-aware intelligent agents can deliver relevant information when a user needs that information. These data make possible many exciting new applications, such as augmented reality, context-aware collaboration, wearable assisted living, augmented manufacturing, and maintenance.

1.1   EXAMPLE: VuMan 3

The VuMan 3 project initiated by Carnegie Mellon University (CMU) provides an example of how the introduction of wearable computing to a task can reap valuable rewards (Figure 1.1).

Many maintenance activities begin with an inspection whose purpose is to identify problems. Job orders and repair instructions are generated from the results of the inspection. The VuMan 3 wearable computer was designed for streamlining Limited Technical Inspections (LTI) of amphibious tractors for the US Marines at Camp Pendleton, California (Smailagic, 1998). The LTI is a 600-item, 50-page checklist that usually takes 4 to 6 hours to complete. The inspection includes an item for each part of the vehicle (e.g., front left track, rear axle, windshield wipers, etc.). VuMan 3 included an electronic version of this checklist. The system's interface was arranged as a menu hierarchy; a physical dial and selection buttons controlled navigation. The top level consisted of a menu that gave a choice of function. Once the user chose an inspection function, the component to be inspected was selected by its location on the vehicle. At each stage, the user could go up one level of the hierarchy.

The inspector selected one of four options for the status of the item: serviceable, unserviceable, missing, or on equipment repair order. The user can also select further explanatory comments about each item (e.g., the part is unserviceable because of four missing bolts).

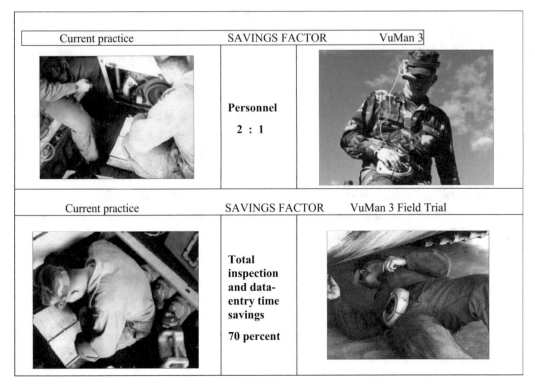

FIGURE 1.1: VuMan 3 maintenance inspection wearable computer. VuMan 3 savings factors.

The LTI checklist contains a number of sections, with approximately 100 items in each section. The user moves through the items by using the dial to select "next item" or "next field." A "smart cursor" helps automate some of the navigation by positioning the user at the most likely next action.

The design process for VuMan 3 included a field study. In typical troubleshooting tasks, one Marine would read the maintenance manual to a second Marine, who performs the inspection. With the VuMan 3, only one Marine is needed for the task as he or she has the electronic maintenance manual with him or her. Thus, the physical manual does not have to be carried into hard-to-reach places.

1.2   VuMan 3: LESSONS LEARNED

The most unanticipated result was a 40% reduction in inspection time. The bottom right image of the VuMan 3 figure demonstrates the reason for this result. Here, the Marine is on his side looking up at the bottom of the amphibious tractor. In such places, it is hard to read or write on the clipboard typically used for inspections. Using a clipboard, the Marine constantly gets into position, crawls out to read instructions, crawls back into position for the inspection, and then crawls out again to record the results. In addition, the Marine tends to do one task at a time when he might have five things he has to inspect in one place. This extra motion has a major impact on the time required to do a task. By making information truly portable, wearable computers can improve the efficiency of this application and many other similar ones.

The second type of time savings with the VuMan 3 occurred when the Marine had finished the inspection. The wearable computer takes a couple minutes to upload its data to the logistics computer. In contrast, the manual process requires a typist to enter the Marine's handwritten text into the computer. Given that the soldier may have written his notes in cold weather while wearing a glove, the writing may require some interpretation. This manual process represents another 30% of total time required for this task.

Such redundant data entry is common when users are performing mobile tasks (Starner, 2004). Numerous checklist-based applications—including plant operations, preflight checkout of aircraft, inventory, and others—could benefit from a form-filling application run on a wearable computer. In the case of the VuMan 3 project, the results were striking. From the inspection start through data entry, the VuMan 3 achieved a 70% time savings compared with the same task done manually. In addition, the wearable computer reduced the maintenance crew from two people to one, and the wearable computer offered a savings in weight over paper manuals.

* * * *

CHAPTER 2

The Wearable Computing UCAMP

Designing wearable computer interfaces requires attention to many different factors because of the computer's closeness to the body and its use while performing other tasks. For the purposes of discussion, we have created the "UCAMP" framework, which consists of the following factors:

- **User**: The user is at the center of the wearable-computer design process.
- **Corporal**: Wearables should be designed to interface physically with the user without discomfort or distraction.
- **Attention**: Interfaces should be designed for the user's divided attention between the physical and virtual worlds.
- **Manipulation**: When mobile, users lose some of the dexterity assumed by desktop interfaces. Controls should be quick to find and simple to manipulate.
- **Perception**: A user's ability to perceive displays, both visual and audio, is also reduced when using a mobile device. Displays should be simple, distinct, and quick to navigate.

Power, heat, on-body and off-body networking, privacy, and many other factors also affect wearable computing (Starner, 2001). Many of these topics are the subjects of current research, and much work and study will be required to examine how these factors interrelate. For reasons of brevity, this lecture concentrates mainly on the CAMP principles and practices.

2.1 USER

Among the most challenging questions facing wearable-computer designers are user needs and interactions. As computing devices move from the desktop to mobile environments, many conventions of human interfacing must be reconsidered for their effectiveness. In addition, mobile users are more impatient than desktop users, expecting their wearable computer to operate more like a flashlight than a rebooting computer. Researchers at CMU built more than 30 generations of wearable computer systems to identify repeated patterns (such as procedures, work orders, team collaboration). Throughout this process, the user has been at the center of wearable-computer design. Based on user interviews and observation of their operation, baseline scenarios are created for current practice. A visionary scenario is created to indicate how technology could improve the current practice and identify opportunities for technology injection. Both scenarios are reviewed with the end user. The user continues to be involved throughout the design, implementation, and testing process. User feedback on scenarios and storyboards become input to the conceptual design phase. Designers alternate between the abstract and the concrete; preliminary sketches are evaluated, new ideas emerge, and more precise drawings are generated. This iterative process continues with soft mock-ups, appearance sketches, and computer and machine shop prototypes, until finally the product is fabricated.

A User-Centered Interdisciplinary Concurrent System Design Methodology (UICSM) (Siewiorek and Smailagic, 1994), based on user-centered design and rapid prototyping, has been applied to the design of wearable computers. Because of UICSM, we have achieved a 4-month design cycle for each new generation of wearable computers. As depicted in Figure 2.1, designers initially visit the user site for a walkthrough of the intended application. A second visit to the user site, after a month of design work, ends the conceptual phase. During this visit, the designers obtain responses from users to storyboards that depict the use of the new device and the information content on the wearable computer's screen. After the second month of design and prototyping, the designers evaluate a software mock-up of the new system running on a previous-generation wearable computer. This represents the results of the detailed design phase. During the third month, the

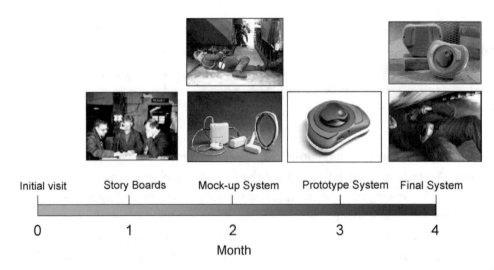

FIGURE 2.1: Four-month design cycle.

design is implemented and users analyze a prototype of the system. The designers deliver the final system after the fourth month of field-trial evaluation.

2.2   CORPORAL

The term *wearable* implies the use of the human body as a support environment for the object described. Society has historically evolved its tools and products into more portable, mobile, and wearable form factors. Clocks, radios, and telephones are examples of this trend. Computers are undergoing a similar evolution. Simply shrinking computing tools from the desktop paradigm to a more portable scale does not take advantage of a whole new context of use, however. Although it is possible to miniaturize keyboards, human evolution has not kept pace by shrinking our fingers. There is no Moore's Law (the number of transistors that fit on a chip doubles every 2 years) for humans.

The human anatomy introduces minimal and maximal dimensions that define the shape of wearable objects, and the mobile context defines dynamic interactions. Conventional methods of interaction, such as keyboard, mouse, joystick, and monitor, have mostly assumed a fixed physical relationship between user and device. With wearable computers, the user's physical context may be constantly changing. Symbol Technologies developed a wearable computer for shipping hubs that provides a good example of how computing must be adapted for the human body.

2.2.1   Example: Symbol Technologies

As a company, Symbol Technologies (now part of Motorola) is well known for its barcode technology. Symbol was also one of the first successful wearable computer companies, having sold more than 100,000 units from its WSS 1000 series (see Figure 2.2). The WSS 1000 consists of a wrist-mounted computer that has a laser barcode scanner encapsulated in a ring worn on the user's finger. This configuration allows the user to scan barcodes while keeping both hands free to manipulate the item being scanned. Because the user no longer has to fumble with a desk-tethered scanner, the Symbol device increases the speed at which the user can manipulate packages and decreases the overall strain on the user's body. Such features are important in shipping hubs, where millions of packages are scanned by hand every year.

Symbol spent more than $5 million and devoted 40,000 hours of testing to develop this new class of device. One of the major challenges during development of the device was adapting the computer technology to the needs of the human body (Stein et al, 1998).

One of the first observations made during the design and testing of this device was that users vary widely in shape and size. Specifically, Symbol's scanner had to fit the fingers of both large

FIGURE 2.2: Symbol's wrist-mounted wearable computer with ring scanner.

men and small women. Similarly, the wrist unit had to be mounted on both large and small wrists. Although the system's wires were designed to be unobtrusive, the device must be designed to break away if entangled and subjected to strain. This design criterion provided a safeguard for the user.

Initial testing revealed other needs that were obvious in hindsight. For example, the system was strapped to the user's forearm while that person was moving boxes. Soon the "soft-goods" materials, which were designed for the comfort of the user, became soaked with sweat. When one user finished the work shift, that person was expected to pass the computer to its operator on the next shift. Not only was the sweat-soaked computer mount considered "gross," it also presented a possible health risk. The designers solved problem by separating the computer mount from the computer itself. Each user received his own mount, which he could adjust as desired. After each shift, the user could remove the computer from the user's mount and give the device to the next user.

Another unexpected discovery was that users tended to use the computer as body armor. For example, when a shipping box would begin to fall, the user would block the box with the computer mounted on his forearm. Similarly, the Symbol designers were surprised to see users adapt their work practices to use the rigid forearm-mounted computer to force boxes into position. To solve these potential problems, the designers made the computer's case of high-impact materials.

Yet another surprise came with longer-term testing of the computer. Employees in the test company's shipping hubs constantly reached into wooden crates to remove boxes. As they reached into the crates, the computer would grind along the side of these crates. Repeated rubbing against the crates wore holes in the computer's casing, eventually damaging the circuitry inside. Therefore, the designers changed the composition of the casing to be resistant to both abrasion and impact.

2.2.2   Symbol's Wrist-Mounted Computer: Lessons Learned

After several design cycles, Symbol presented the finished system to new employees in a shipping hub. After a couple of weeks' work with the new system, test results showed that the new employees felt the system was cumbersome, whereas established employees who had participated in the design of the project felt that the wearable computer provided a considerable improvement over the old system of package scanning. After consideration, Symbol's engineers realized that these new employees had no experiential basis for comparing the new system to the past requirements of the job.

To better acclimate new employees to the new system, Symbol began its training of new shipping-hub employees with 2 weeks of training in which the employees were taught their job using the old system of package scanning—the employee would reach into a crate, grasp a package, transfer it to a table, grasp a handheld scanner, scan the package, replace the scanner, grasp the package, and transfer it to the appropriate conveyer belt. The employees were then introduced to the forearm-mounted WS-1000. With the wearable computer, the employee would squeeze together his index and middle finger to trigger the ring-mounted scanner to scan the package while reaching for it, then grasp the package, and transfer it to the appropriate conveyor belt in one fluid motion. These new employees gave very positive scores for the wearable computer.

This lesson, that perceived value and comfort of a wearable computer are relative, was also investigated by Bodine and Gemperle (2003). In short interviews, users were fitted with a backpack or armband "wearable" and told that the system was a police monitoring device (similar to those used for house arrest), a medical device for monitoring health, or a device for use during parties. The subjects were then asked to rate the devices on various scales of desirability and comfort. Not surprisingly, the police wearable was considered the least desirable. In addition, the police wearable elicited more negative physical comfort ratings and the medical wearable elicited more positive physical comfort ratings although they were the same device. In other words, perceived comfort can be affected by the supposed function of the device.

Plan Ahead: Field tests often have to be set up months in advance to accommodate people's schedules as well to allow time for paperwork and clearances. Sometimes a designer team must race to complete the prototype units before the start of the field tests. While our team was developing Navigator 2, which was designed for sheet metal inspection of KC-135 aircraft at McClellan Air Force Base in Sacramento, CA, we encountered a debugging problem. The boards arrived on Friday, and we hand-assembled one to ensure that the physical circuit worked. The field test was scheduled to begin the following Tuesday, and the team had made flight reservations for Sunday evening.

We tested the board and found that although the clocking circuit on the board appeared to be simple, we could not get the microprocessor clock to stay on. We could see the clock start, but within a couple of milliseconds it would turn off. We tried many different hardware configurations in an attempt to get the clock to stay on. Then a member of the design team had an idea that the problem might actually be in the software. Examination of the software showed a loop that was to determine whether the power-on switch had been set. Because switches sometimes bounce up and down, the purpose of the loop was to wait until the bouncing had stopped so that a single button press would not be interpreted as multiple depressions. One push of the button was to start the system by turning the processor clock on, a second push would turn the system off by turning off the processor clock. It turned out that a design error in the software had the clock turn-off instruction inside the loop rather than outside. Thus, every time the switch bounced, the clock would be turned off. Once we found the problem, we provided a remedy easily, but we worked through the night and had to delay the departure of the field test crew until noon Monday. The crew finished assembling the two units in their motel room Monday evening, and the test started on time, but just barely.

2.2.3   Wearability Design Criteria

Researchers have also explored wearability in more general terms. Wearability is defined as the interaction between the human body and the wearable object. Dynamic wearability includes the human body in motion. Design for wearability considers the physical shape of objects and their active relationship with the human form. Gemperle and coauthors explored history and cultures, including topics such as clothing, costumes, protective wearables, and carried devices (Gemperle et al., 1998; Siewiorek, 2002). They studied physiology, biomechanics, and the movements of modern dancers and athletes. Drawing on the experience of CMU's wearable computing group of more than 30 generations of machines representing more than a hundred person years of research, they codified the results into guidelines for designing wearable systems. These results are summarized in Table 2.1.

The Gemperle team also developed a set of wearable forms to demonstrate how wearable computers might be mounted on the body. Designers also planned the pods so that they could house electronic components. Each form follows wearability design guidelines. All of the forms are between 3/8 and 1 inches thick; flexible circuits can fit comfortably into the 1/4-inch-thick flex zones.

Beginning with acceptable areas and the humanistic form language, the team considered human movement in each individual area. Each area of the body is unique, and some study of the muscle and bone structure was required along with study of common movements. Perception of size was studied for each individual area. For testing, minimal amounts of spandex were stretched around the body to attach the forms. Figure 2.3 shows the forms.

TABLE 2.1: Design criteria for wearability	
ATTRIBUTE	**COMMENTS**
Placement	Identify where the computer should be placed on the body. Issues include identifying areas of similar size across a varied population, areas of low movement or flexibility, and large surface areas.
Humanistic form language	The form of the object should work with the dynamic human form to ensure a comfortable fit. Principles include making the inside surface concave to fit body, making the outside surface convex to deflect objects, tapering sides to stabilize the form on the body, and radiusing edges or corners to provide soft form.
Human movement	Many elements make up a single human movement: mechanics of joints, shifting of flesh, and flexing and extending of muscles and tendons beneath the skin. Allowing for freedom of movement can be accomplished in one of two ways: by designing around the more active areas of the joints or by creating spaces on the wearable form in which the body can move.
Human perception of size	The brain perceives an aura around the body. Forms should stay within the wearer's intimate space so that perceptually they become a part of the body. The intimate space is between 0 and 5 inches off the body and varies according to position on the body.
Size variations	Wearables must be designed to fit many types of users. Allowance for size variations is achieved in two ways: static anthropometric data, which details point-to-point distances on different-sized bodies, and consideration of human muscle and fat growth in three dimensions using solid rigid areas coupled with flexible areas.
Attachment	Comfortable attachment of forms can be created by wrapping the form around the body rather than using single-point fastening systems such as clips or shoulder straps.
Contents	The system must have sufficient volume to house electronics, batteries, and other components; this, in turn, constrains the outer form of the device.
Weight	The weight of a wearable should not hinder the body's movement or balance. The bulk of the wearable object's weight should be close to the center of gravity of the human body, thereby minimizing the weight that spreads to the extremities.

	TABLE 2.1: (*continued*)	
ATTRIBUTE	**COMMENTS**
Accessibility	Before finalizing design of a wearable system or manufacturing it, walk and move with the wearable object to test its comfort and accessibility.
Interaction	Passive and active sensory interaction with the wearable object should be simple and intuitive.
Thermal	The body needs to breathe and is very sensitive to products that create, focus, or trap heat.
Aesthetics	Culture and context will dictate shapes, materials, textures, and colors that perceptually fit the user and their environment.

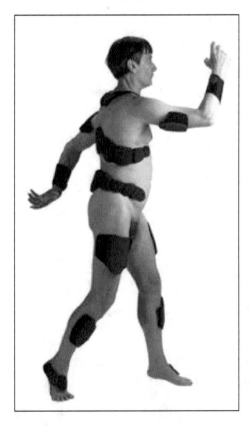

FIGURE 2.3: Forms studied for wearability.

A Hot August Day in Sacramento: Throughout our design of wearable computers, we have been sensitive to the amount of energy consumption and the heat generated during operation of the units. The Navigator 2 system was designed for sheet metal inspection of KC-135 aircraft. Although it only weighed 4 lb with batteries, Navigator 2 was heavier than our guidelines for wearing on the hip. The user wore an 8-inch-wide belt, which held Navigator 2 in a pouch that fit in the small of his back. Although the heat generated by Navigator 2 was negligible, we received complaints from aircraft inspectors using the system at McClellan Air Force Base in Sacramento, CA, in August. The 8-inch-wide belt acted as a vapor barrier, preventing the users' perspiration from evaporating and thus cooling their skin. Thereafter, we began to consider the footprint of our systems, in addition to weight and heat generation.

These studies and guidelines provide a starting point for wearable systems designers. However, there is still much work to be done in this area. For example, these studies did not consider the weight of wearable devices nor did they examine the long-term physiological effects such systems might have on the wearer's body. Similarly, fashion can affect the perception of comfort and desirability of a wearable component. As wearable systems become more common and are used for longer periods, it will be important to test these components of wearability.

Does this Bend? Fort Gordon in Georgia has an Army Battle Lab, which tests equipment for applicability and ruggedness. All of the CMU wearable computers have been rugged enough for field evaluations, and VuMan 2, because of its simplicity, has performed reliably for more than 8 years. When we took VuMan 2 to the lab, a staff sergeant grabbed the unit's PCMCIA memory card (about the size of a credit card and about five times as thick) and visibly flexed the card, explaining that most electronics were not rugged enough for field use. We were sure the card had been bent beyond its specified limits, and we held our breath as we inserted the card to see if the system still worked. It did!

Example: The MICON (Microprocessor CONfiguration) Project. Sometimes the confluence of several independent events leads to a major change in direction. In 1981, Raj Reddy, then the head of the CMU Robotics Institute, asked one of the authors if there was a way to expedite design of single-board computers for robotic manipulators. The Robotics Institute, founded in 1979, was pioneering the applications of robotic manipulators and integrating sensors to improve their capability. However, each new concept required new capabilities from the single-board computer controlling the manipulation; often these capabilities were beyond what was commercially available. Yet the improvements needed were relatively simple, straightforward extensions such as doubling the size of memory. But the design process was laborious: a technician would have to become intimately familiar with the processor and memory chips used, create new logic to glue these chips together, synthesize a design, create a parts list and net list, send the design out to be fabricated, and debug the design once the printed circuit board had been returned. This process often took 6 months. Raj's question was, "Why couldn't this be done in 24 hours?"

That was the beginning of the MICON project, which involved a number of researchers and resulted in a system that could generate a design from high-level specifications in less than 10 min. (Over its life, this project also produced about half a dozen Ph.D.'s, a dozen master's degrees, and two dozen undergraduate projects.) The specifications needed were similar to information found on the back page of a product brochure, the specific processor and its clock frequency, amount of memory, type of input/output devices, thermal heat dissipation, power requirements, board area, and cost. From these specifications, MICON would select parts from its database and fill in the details for gluing the parts together. For example, an appropriate clocking circuit would be developed to drive the microprocessor chip. Terminating resistors would automatically be added to on-board buses. Ultimately, in its commercial version, MICON could produce an initial design in a couple of minutes and go into an optimization phase that would produce a thousand designs per minute, so that by the end of about 10 min, a design consuming half the power and costing half as much as the original design would be synthesized. To demonstrate MICON's capability, we started producing and fabricating designs in 1988.

During the summer of 1990, the authors taught a course to 25 midcareer professionals on design of electromechanical systems. Rather than spend the 12-week course focusing on theory, we decided to put theory into practice by designing and fabricating a single-board computer system. One of the undergraduate students working on the project discovered the Private Eye Head-Mounted Display, which led to prototyping our first wearable computer in the fall of 1990. The team also coined the term *wearable computer*, which come to define this new class of computing. The first application was reading maps and architectural blueprints.

Early Wearable Designs: Lessons Learned. The VuMan 1 design team consisted of an undergraduate industrial designer and graduate students for electronics, mechanical engineering, and software design. The team quickly learned that design mistakes often occur at the boundaries between disciplines where one discipline takes something for granted and does not communicate it to a second discipline.

For example, every good electrical engineer designs in a reset button so that if the electronics gets confused, the device can be set to a known state. The team built 30 boards with bright red reset buttons. Upon seeing the boards, the industrial designer pointed to the red button and asked, "What's that?" The electrical engineer responded, "Why, a reset button, of course." But there was no corresponding hole for the reset button in the 30 housings the team had already fabricated. The two designers had not discussed the reset button in advance.

The mechanical engineer and industrial designer quickly drilled holes in 30 housings for the reset button. Because the reset button had not been part of the mechanical design, however, there was no room between the housing and the printed circuit board for the reset button. So all 30 printed circuit boards had to be modified to move the reset button further from the edge of the board. However, moving the reset button caused the button's metal contact to short out pins on the memory chip. The team added some electrical tape to provide insulation between the reset button pins and the memory chip pins.

In this situation, one miscommunication about the necessity for a reset button cascaded into a series of reworks. After the design for VuMan 1 was complete, the team did a careful analysis of the design methodology, which indicated that the miscommunications among designers (which we call "design escapes") resulted in approximately 25% more time and cost. Since that early experience, the team has developed a concurrent interdisciplinary design methodology; this allows design to proceed concurrently among the disciplines and ensures timely communication of design decisions across disciplinary boundaries.

2.3   ATTENTION

Humans have a finite capacity that limits the number of concurrent activities they can perform, which presents a special challenge to designers of wearable computers. Nobel laureate Herb Simon observed that human effectiveness is reduced as a person tries to multiplex more activities. Frequent interruptions require a refocusing of attention. After each refocus of attention, a period of time is required to reestablish the context before the interruption. In addition, human short-term memory can hold seven plus or minus two (i.e., five to nine) chunks of information. With this limited capacity, today's systems can overwhelm users with data, leading to information overload. The challenge for human–computer interaction design is to use advances in technology to preserve human attention and to avoid information saturation.

In the mobile context, the user's attention is divided between the computing task and activities in the physical environs. Some interfaces, such as some augmented realities (Azuma, 1997) and dual-purpose speech (Lyons et al., 2004a), try to integrate the computing task with the user's behavior in the physical world. The VuMan 3 interface did not tightly couple the virtual and real worlds, but the computer interface was designed specifically for the user's task and allowed the user to switch rapidly between a virtual interface and the hands-on vehicle inspection.

Yet many office productivity tasks, such as e-mail or Web searching, have little relation to the user's environment. The mobile user must continually assess what attentional resources he can commit to the interface and for how long before switching his attention back to his primary task. Oulasvirta et al. (2005) specifically examined such situations by fitting cameras to mobile phones and observing users attempting Web search tasks while following predescribed routes. Participants performed these tasks in a laboratory, in a subway car, riding a bus, waiting at a subway station, walking on a quiet street, riding an escalator, eating at a cafeteria and conversing, and navigating a busy street. Web pages required an average of 16.2 seconds to load and had considerable variance, requiring the user to pay attention to the interface. The degree to which subjects shifted their attention away from the phone interface depended on the environment; participants shifted their attention away during 35% of page loadings in the laboratory and 80% of page loadings while walking a quiet street. The duration of continuous attention on the mobile device also varied according to the physical environment; attention was 8 to 16 seconds for the laboratory and cafe compared to less than 6 seconds for while riding the escalator or navigating a busy street. Similarly, the number of attention shifts depended on the demands of the environment.

The authors note that even riding an escalator demands attention (such as choosing a correct standing position, monitoring personal space for passersby, or determining when to step off). Accordingly, they are working on a Resource Competition Framework, based on the multiple resource theory of attention (Wickens, 2000), to relate mobile task demands to the user's cognitive resources. This framework helps predict when the mobile user will need to adopt attentional strategies to cope with the demands of a mobile task.

The Oulasvirta team observed four such strategies in their study. The first, calibrating attention, refers to the process whereby the mobile user first attends to the environment and determines the amount of attention he needs to devote to the environment versus the device's interface. A second strategy, brief sampling over long intervals, refers to the practice of only attending to the environment in occasional brief bursts to monitor for changes that may require a deviation from plan, such as when reading while walking along an empty street. The third strategy, task finalization, refers to subjects' preference to finish, when sufficiently close, a task or subtask before switching attention back to the physical environment. The fourth strategy, turn-taking capture, occurs when the user is conversing with another person. Attending and responding to another person requires significant concentration, leading to little or no attention to the mobile interface.

Thad Starner, coauthor of this lecture, has been using a wearable computer to take notes on daily life since 1993. He has noticed similar strategies in his interactions while using the wearable device. Describing these attentional strategies more fully and designing interfaces that leverage them will be important for developing future mobile interfaces. Much research has been performed on aircraft and automobile cockpit design-to-design interfaces that augment but do not interfere with the pilot's primary task of navigating the vehicle. However, only recently have we begun to observe mobile users closely and examine interface use (and misuse) in the field. With these added observation techniques, we can apply and test theories of attention in everyday situations.

This newfound ability to monitor mobile workers may help us determine how not to design interfaces. In contrast to the VuMan 3 success described above, Ockerman's (2000) work, "Task Guidance and Procedure Context: Aiding Workers in Appropriate Procedure Following," warns that mobile interfaces may hinder the user's primary task if they are not properly designed. This project studied experienced pilots who were inspecting their small aircraft before flying. When a wearable computer was introduced as an aid to completing the aircraft's safety-inspection checklist, the expert pilots touched the aircraft less. (Many pilots develop an intuition about the aircraft's condition by touching the plane.) In addition, the pilots relied too much on the wearable computer system, which was purposely designed to neglect certain safety steps. The pilots assumed that the wearable computer checklist was complete instead of relying on their own mental checklists. Ockerman shows how such interfaces might be improved by providing more context for each step in the procedure. That is, the higher the fidelity of the computer information, the more attention the user gives to the computer (as opposed to the physical environment). This study used three different versions of information: text, text with sketches, and text with photographs. Another approach would be integrating the aircraft itself into the interface (e.g., use augmented reality to overlay graphics on the aircraft indicating where the pilot physically inspects the plane).

Most recently, the Defense Advanced Research Projects Agency's Augmented Cognition project (Kollmorgen et al., 2005) aims to create mobile systems that monitor their users' attentional resources and records or delays incoming information so that the user receives information in a more orderly and manageable time sequence. These systems use mobile electroencephalogram readings or functional near-infrared imaging to monitor the user's brain activations and relate these results to the user's current state (Archinoetics, http://www.archinoetics.com). Such projects, if successful on a larger scale, could reveal much about the mental resources required for truly mobile computing.

The Attention Matrix (Anhalt et al., 2001), shown in Table 2.2, categorizes activities by the amount of attention they require. The activities are information, communication, and creation. Individual activities are categorized by the amount of distraction they introduce, measured in units of increasing time: snap, pause, tangent, and extended. The snap is an activity that is usually completed in a few seconds, such as checking your watch for the time. The user should not have to interrupt his primary activity to do this. The pause requires the user to stop his current activity, switch to the new but related activity, and then return to his previous task within a few minutes. Pulling over to

TABLE 2.2: Attention matrix

Time ⟶

	Snap	Pause	Tangent	Extended
Information				
active — Receiving, Notifying, Monitoring, Serendipity	Message arrival, Information accessible, Auction, Stocks, Sports, Matching similar needs, Free food			– Audio, Walkman – Transferring files from network – Reading news
— Seeking	Line length, Bus arrival, Locate person	– Exam calendar – Software/hardware help – Calendaring – Navigation	– Looking for Class Notes – Who else is doing this now? – Access personal data	
passive — Browsing, Finding		Information on web or built environment	– Poster, bulletin board information	– Web Research – Reviewing Class Notes
— Verifying		– Recall previous queries – Double checking information		
Communication				
artificial — Initiating	S.O.S. Emergency	– Introductions	– Team building – Collaborative work – Event planning – Assassins game – Social Planning	– Chatting (public or private)
— Participating	Instant messaging	– Queries		
informal — Broadcasting		– Information exchange – Scheduling	– Posting information to bulletin board – Advertising	
formal • One to One communications with an individual • One to Group communications with select group, team or family • One to All Possible broadcast communications with unknown people				
Creation				
work — Recording	Remember this!, Add a todo or call list		– Class note taking – Meeting – Filling out survey – Registration – New ideas – Adding information to existing projects	– Generating messages
— Synthesizing		Forwarding x to y		– Summarizing lecture
— Generating				– Mobile tool building

the side of the road and checking directions is an example of a pause. A tangent action is a medium-length task that is unrelated to the action the user is engaged in. Receiving an unrelated phone call is an example of a tangent activity. An extended action is when the user deliberately switches tasks, beginning a wholly new long-term activity. For the car driver, stopping at a motel and resting for the night is an extended activity.

Because distractions on the left of the matrix take less time from the user's primary activity, our intent is to move activities of the matrix toward the left side (snap). Our goal is to evaluate how this process extends to a larger sample of applications.

2.3.1   Example: VuMan 3's Attention-Based Interface

VuMan 3 (see Figure 1.1) provides a simple interface that is designed for operations when the user's physical attention is occupied, such as during inspection of Marine vehicles. The VuMan3 has a low-resolution display and, consequently, a purely textual interface. Figure 2.4 shows a sample screen from the user interface.

The user navigates through a geographically organized hierarchy: top, bottom, front, rear; then left, right, and more detail. Further down the hierarchy, at the node leafs, individual components are identified. (There are more than 600 of these components.) The VuMan 3 user indicates each component as "serviceable" or "unserviceable." If the component is serviceable, no further information appears on the screen. If the component is unserviceable, the next screen displays a short list of possible reasons.

The user can move up the hierarchy (by choosing the category name in the upper right corner) or move to the next component. Once a component is marked as serviceable or unserviceable, the next item in the sequence is automatically displayed for the user. Each component has a probability associated with it of being serviceable or unserviceable, and the cursor is positioned over the most likely response for that component. (VuMan 3 has three types of buttons that all performed the same function. The button locations are designed for left-hand and right-hand thumb-dominant users and for finger-dominant users.)

The VuMan 3 screen contains navigational information. Sometimes there is more on a logical screen than can fit on a physical screen, so the designer must provide easy manipulation among the physical screens that display this logical information. Navigation icons are on the left side of the screen. The user can go to the previous screen or next screen (which are both functional parts of the logical screen). In addition, the user can always go back to the main menu (Figure 2.5).

In Figure 2.5, the options include inspection information (the vehicle name and serial number, the number of hours the vehicle has been used, and other details that distinguish this inspection report from other reports) and the inspection list. The inspection checklist is divided into sections, and different people can be inspecting different sections in parallel. An inspector would pick a section,

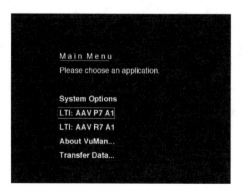

FIGURE 2.4: VuMan 3 options screen.

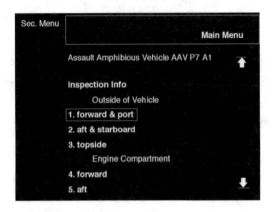

FIGURE 2.5: VuMan 3 "information" screen.

highlight it by rotating the VuMan 3's dial ("1. forward & port" in Figure 2.5), and then select the highlighted item by pressing a button (clicking). The inspector would then receive a detailed set of instructions on what to do.

In Figure 2.6, the inspector is instructed to check for damage and bare metal. The "smart cursor" anticipates that the inspector will be filling in the status field whose current value is "none." When the inspector selects a component (using the button on VuMan 3), a list of options appears; the first is serviceable. In this type of inspection, the item is serviceable in 80% of the cases. By putting the most probable selection first, the interface emulates a paper checklist where most of the items will be checked as "OK." The smart cursor then assumes the most likely next step. The inspector does not need to move the dial; he merely clicks on the highlighted option.

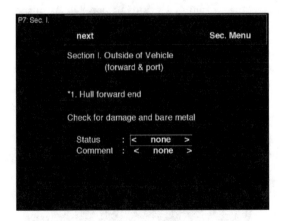

FIGURE 2.6: VuMan 3 "hull forward" screen.

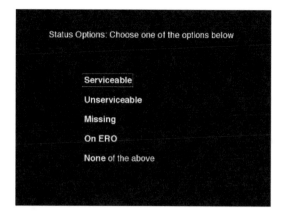

FIGURE 2.7: VuMan 3 "Status" screen.

Figure 2.7 shows an example of the next likely screen; "serviceable" is selected. If the inspector chooses serviceable for all components, he would simply tap the VuMan 3 button multiple times. If an item is unserviceable, he would turn the dial to select "unserviceable." Then he chooses options form a list of reasons why that particular device is unserviceable. The inspector can select one or more reasons for unserviceability; he selects "done" when all reasons have been entered. The selected reasons appear in the comment field. When the inspector completes work on the checklist, the data are uploaded to the logistics computer, which then generates the work orders for components needing service or replacement.

2.4   MANIPULATION

As noted in the example above, VuMan 3 added a novel manipulation interface suitable for use when physical attention is occupied. This device is a good example of wearable-computer design that addresses demands for the user's attention and offers simple manipulation within a complex set of information. (Figure 1.1 shows two views of VuMan 3 in use.)

2.4.1   VuMan 3 Manipulation Design: Lessons Learned

Designing and building the VuMan 3 system has provided several lessons to the CMU team. As part of the design cycle, the designers tested a mouse (essentially a disk with buttons). The physical configuration of the mouse disk could be ambiguous, however. Was the left button in the proper position when the mouse's tail was toward the user or away from the user? Were the buttons supposed to be at the top? The dial removed this orientation ambiguity.

Another design lesson was the need to minimize cables. An earlier system had cables connecting to a separate battery as well as to the mouse and the display. These wires quickly became knotted. Internal batteries were used in the VuMan 3 design to avoid this problem, and the dial was built into the housing. The only remaining wire was the one to the display.

A third design lesson was that wearable computers have a minimum footprint that is comfortable for a person's hand. Although the keyboards of palmtop computers are getting smaller, evolution has not correspondingly shrunk our fingers. Future electronics designs will be thinner—eventually no thicker than a sheet of plastic or a piece of clothing. But such devices will need a minimal footprint for their interface.

Furthermore, the interface—no matter where it is located on a person's body—is operated in the same way. This consistent use is a major feature of the dial. It can be worn on the hip or in the small of the back. In airplane manufacturing, where workers navigate small spaces, the hip defines the smallest diameter through which the person can enter. In these situations, a shoulder holster is preferred for a wearable computer. For VuMan 3, the Marine's oversized coverall pockets offered another option for the location of the wearable computer. The soldiers could drop the computer into their coveralls and operate it through the cloth of the pocket.

For simplicity, as well as orientation independence, the dial integrated well with the presentation of information on the screen. All information on the screen could be considered to be on a circular list. In most cases, VuMan 3 shows fewer than a dozen selectable items on a screen. This sparse screen is a particular advantage on a head-mounted display, where the user may be reading while moving. In such devices, the font must be large enough to read while the screen is bouncing.

The dial could also be an intuitive interface for Web browsing. Probably, there are no more than a half-dozen items to select on a typical Web page; the user can rotate the dial clockwise or counterclockwise for quick navigation then use a button to select the highlighted item.

2.4.2   Other Manipulation Devices

A number of other manipulation devices have been developed and tested for wearable computing. Three that are especially suitable for wearable computers include a wheel-pointer, a specialized dial, and a small keypad or keyboard.

CMU Wheel-Pointer. The wheel-pointer is a small, lightweight device that provides a very flexible interaction. Designers created this device when they removed the dial from a wearable computer and put it into a separate module. A user can operate the wheel-pointer while holding it in his hand, or he can wear the device on the body or attach it to clothing (Figure 2.8). When worn, the device

FIGURE 2.8: CMU wheel-pointer.

can be attached by variety of methods, such as a belt or hook-and-loop fabric, or the user can place it in a pocket or pouch.

The wheel-pointer was designed with wearability in mind, with an emphasis on use in harsh environments. This design focus makes the wheel-pointer especially well suited for mobile and wearable applications in industrial and military settings, as well as traditional consumer and business environments.

The wheel-pointer operates the same way as the VuMan 3's dial, with the addition of a center pad that controls a cursor. The user turns the dial (wheel) to select an option, then presses the dial to enter that choice.

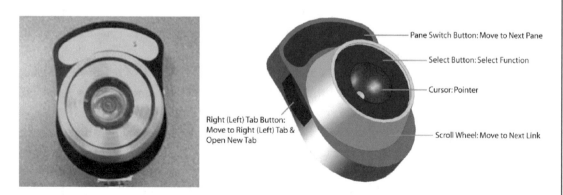

FIGURE 2.9: MoRE multifunction dial.

MoRE Multifunction Dial. The MoRE multifunction dial's components include a wheel, select button, pane-switch button, two tabbing buttons, and pointer knob, as shown in Figure 2.9. The dominant feature of the dial is the wheel.

By turning the wheel clockwise or counterclockwise, the user can move the selection box to the next or previous link. The wheel scrolls an onscreen selection box up and down the page; it displays more information when the selection box reaches the top or bottom of the screen (but not the actual top or bottom of the pane, which is larger than the screen). When the selection box reaches the top or bottom link on the pane, continuously turning the wheel will move the selection box back to the bottom or top link of the pane. The inner area of the ring of the wheel acts as its button. The center knob within the button has the capability to move the cursor on the screen in eight-direction movement, similar to using a mouse or other pointer. The large button placed on the upper section of the dial is the pane-switch button. By pressing this button, the user can move the selection box to the next pane of information in the current tab. Pressing the left or right tab button will open the tab accordingly.

The dial interacts with the user interface of the manual integration software. The software provides four panes in which contents of the manuals, options for the screen, results of a user's search queries, and annotation-recording interface are presented. Each component of the user interface and the dial is functionally connected to support the application (Smailagic and Siewiorek, 2007).

2.4.3   Mobile Keyboards

The VuMan 3 addressed the problem of menu selection in the mobile domain and effectively used a one-dimensional dial to create a pointing device that can be used in many different mobile domains. For tasks such as wireless messaging, however, more free-form text entry is needed. Although speech technology has made great strides in the last decade, speech recognition is very difficult in the mobile environment and often suffers from high error rates. In addition, speech is often not socially acceptable (e.g., in hospitals, meetings, or classrooms). Keyboard interfaces still provide one of the most accurate and reliable methods of text entry.

Since 2000, wireless messaging has been creating billions of dollars of revenue for mobile phone service providers, and more than 1 trillion messages are currently being typed per year. Until recently, many of these messages were created using the multitap or T9 input method on phone keypads, in which nine keys are used to enter all text (often by tapping a key multiple times to enter a character). Yet studies have shown that users average a slow 10 to 20 words per minute (wpm) using these common typing methods; for comparison, a highly skilled secretary on a desktop keyboard averages 70 to 90 wpm. Given the obvious desire for mobile text input, human–computer interface researchers have begun reexamining keyboards. Although standard keyboard entry has been well studied in the past, mobility suggests intriguing possibilities. For example, if an adequate method of typing can be combined with a sufficient display for the mobile market, computing may move off the desktop permanently.

Traditionally, text-entry studies emphasize learnability, speed, and accuracy. But a mobile user may not be able to devote his full attention to the text-entry process. For example, he may be taking notes on a conversation and wish to maintain eye contact with his conversational partner. Or he may be in a meeting and use his keyboard under the desk to avoid distracting others with the keyboard's noise and the motion of his fingers. A user might also attempt to enter text while walking and need to pay attention to his physical environment instead of looking at the screen. These conditions all describe "blind" typing, in which the user enters text while only occasional glancing at the screen to ensure that the text has been entered correctly.

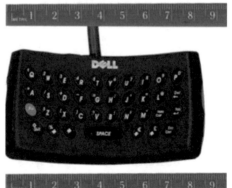

a. Handy key's Twiddler   b. Mini-QWERTY thumb keyboard

FIGURE 2.10: (a) Handykey's Twiddler. (b) Mini-QWERTY thumb keyboard.

Lyons et al. (2004b) and Clawson et al. (2005) have performed longitudinal studies on two keyboards, Handykey's Twiddler (Figure 2.10a) and the mini-QWERTY thumb keyboard (Figure 2.10b), to determine if these types of devices might achieve desktop-level text entry in the mobile domain. As the average desktop text-entry rate was considered to be 30 wpm, including hunt-and-peck typists, this benchmark was chosen as the minimum for speed. Traditionally, very high accuracy is desired for desktop typing. However, as a culture of informal e-mail and text messaging has developed, less accurate typing has become common. The research community is debating how to reconcile speed and accuracy measures; at present, error rates of approximately 5% per character are common in current mobile keyboard studies.

Twiddler. With the Twiddler, novices averaged 4 wpm during the first 20-minute session and 47 wpm after 25 hours of practice (during 75 separate 20-minute sessions). The fastest user averaged 67 wpm, which is approximately the speed of one researcher who has been using the Twiddler for 12 years. Although 25 hours of practice seems extreme, a normal high school typing class involves almost three times that much practice to achieve a goal of 40 wpm.

Mini-QWERTY. Many mobile computer users may already have experience with desktop QWERTY keyboards. Because they are familiar with the key layout, these users might more readily adopt a mini-QWERTY keyboard for mobile use. Can a mini-QWERTY keyboard achieve desktop rates? The study performed by the Clawson team examined the speed and accuracy of experienced desktop typists on two different mini-QWERTY keyboards. These users averaged 30 wpm during the first 20-minute session and increased to 60 wpm by the end of 400 minutes of practice!

Although participants in both studies easily achieved desktop typing rates and had error rates comparable to those in past studies, can these keyboards be used in mobile environments? Neither study tested keyboard use while the user was walking or riding in a car, but both research teams experimented with blind text entry (in at least one condition, typists could not look at the keyboard or the see output of their typing). When there was a statistically significant difference between blind and normal typing conditions, experienced Twiddler typists had slightly improved speeds and decreased error rates. In contrast, experienced mini-QWERTY typists were significantly inhibited by the blind condition, with speeds of 46 wpm and approximately three times the error rate even after 100 minutes of practice. These results might be expected, in that Twiddler users are trained to type without visual feedback from the keyboard whereas the mini-QWERTY keyboard design assumes that the user can see the keyboard to help distinguish among the horizontal rows of small keys.

The results of these keyboard studies demonstrate that desktop typing rates can be achieved on a mobile device in various ways. The question remains, however, as to whether the benefits of typing quickly while blind or moving will be sufficient to cause users to learn a new text-entry method.

Additional benefits might affect the adoption of keyboards in the future. For example, a 12-button device like the Twiddler can be the size of a small mobile phone and still perform well, whereas a 40-button mini-QWERTY keyboard may already be as small as possible for users' hands.

Another factor may be adoption of mobile computing in developing countries. According to the Techweb online information site, nearly a billion mobile phones were shipped in 2005. Many new mobile phone users will not have learned to type on a Roman alphabet (QWERTY) keyboard, and these users may be more concerned with quick learning than compatibility with desktop input skills.

2.4.4   Speech Interfaces

Vocollect. Mobile keyboards are not suitable for applications in which hands-free control is necessary, such as warehouse tasks. Pittsburgh-based Vocollect focuses on package manipulation—in particular, the warehouse-picking problem. In this scenario, a customer places an order consisting of several different items stored in a supplier's warehouse. The order transmits from the warehouse's computer to an employee's wearable computer (Figure 2.11). The computer then speaks the name and location of each item to the employee through a pair of headphones. The employee can control how this list is announced through feedback via speech recognition, and he can also report inventory errors as they occur. The employee gathers the customer's order from the warehouse's shelves and ships it.

FIGURE 2.11: Vocollect's audio-based wearable computer.

This audio-only interface also frees the employee to manipulate packages with both hands, whereas a pen-based system would be considerably more awkward. As of December 2000, Vocollect had approximately 15,000 users and revenues between $10 and $25 million.

Example: Navigator Wearable Computer with Speech Input. Boeing has been pioneering "augmented reality," using a head-mounted, see-through display. As the user looks at the aircraft, the next manufacturing step is superimposed on the appropriate portion of the aircraft, as depicted on the display screen (Mizell, 2001). The Navigator 2, ca. 1995, is designed for voice-controlled aircraft inspection (Siewiorek et al., 1994; Smailagic and Siewiorek, 1996). The speech-recognition system, with a secondary manually controlled cursor, offers complete control over the application in a hands-free manner, allowing the operator to perform an inspection with minimal interference from the wearable system. Entire aircraft manuals, or portions of them, can be loaded into the Navigator 2 as needed, using wireless

FIGURE 2.12: Navigator 2 in use.

communication. The results of inspection can be downloaded to a maintenance computer. To maximize usability, each item or control may be selected simply by speaking its name. Figure 2.12 shows the Navigator 2 systems in use.

One of the first applications for the Navigator is fabrication of wire harnesses, which differ for each configuration of a plane. All orders for an aircraft are essentially unique; they may be from different airlines or for use on long-haul routes or short-haul routes. The airline may also specify different configurations of the interior. For example, the galleys will be in different places on different planes, the seating would change, and so forth.

Wire harnesses are fabricated months before they are assembled into the aircraft. The assembly worker starts with a pegboard measuring about 3 feet high and 6 feet long. Mounted on the board is a full-sized diagram of the final wire harness. Pegs provide support for the bundles of wire as the worker forms them. The worker selects a precut wire, reads its identification number, looks up the wire number on a paper list to find the starting coordinates of the wire, searches for the wire on the diagram, and threads the wire following the route on the diagram. With augmented reality, using the Navigator, the worker selects a wire and reads the wire identification from its bar code. A head tracker provides the computer with information on where the worker is looking and superimposes the route for that particular wire on the board on the Navigator's display. Trial evaluations of Navigator use indicate a savings of 25% of assembly effort, primarily resulting from elimination of cross-referencing the wire with paper lists.

Another part of Navigator 2's application, three-dimensional inspections, is instructive as well. This application was developed for use at McClellan Air Force Base in Sacramento, CA, to inspect KC-135 aerial refueling tankers. Every 5 years, these aircraft are stripped down to bare metal. The inspectors use magnifying glasses and pocket knives to hunt for corrosion and cracks. The information screen on the Navigator 2 provides options that the inspector selects with voice input.

At startup, the application prompts the user whether to activate the speech-recognition system. The user then proceeds to the main menu. From this location, several options are available, including online documentation, assistance, and the inspection task (Figure 2.13).

From this location, several options are available, including online documentation, assistance, and the inspection task. Once the user chooses to begin an inspection, he specifies the type of plan on an aircraft-selection screen (Figure 2.14).

Next, the inspector narrows the field of interest for the inspection, choosing from major features (left wing, right tail, etc., as shown in Figure 2.15).

Then the inspector proceeds to more specific details (individual panes in the cockpit window glass, as shown in Figure 2.16).

A coordinate system is superimposed on the inspection region of the Navigator 2 display. The horizontal coordinates begin from the nose, and the vertical coordinates are "water lines" derived as if the airplane were floating. The inspector records each imperfection in the skin at the correspond-

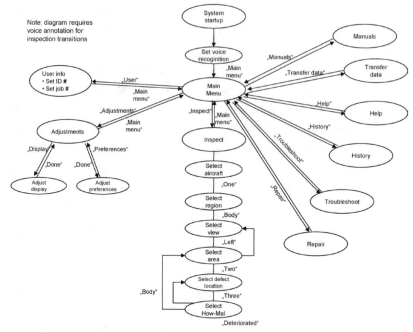

FIGURE 2.13: Navigator 2's application state diagram.

ing location on the display. The inspector records the area covered by each defect, as well as the type of defect—such as corroded, cracked, or missing. To maximize usability, each item or control may be selected simply by speaking its name.

The user navigates to the display corresponding to the portion of the skin currently being inspected. This navigation is partly text-based on buttons (choose aircraft type to be inspected) and partly graphical-based on side perspectives of the aircraft (choose area of aircraft currently being inspected). The inspector can navigate either through a joystick input device or through speech input (speaking exactly the text to be selected). The positioning of the imperfection is done solely through the joystick because speech is not well suited for pointing to indicate the position of the imperfection. As the inspector moves the cursor with the joystick, the coordinates and the type of material at the cursor location are displayed at the bottom of the screen. If the inspector sees a defect at the current position, a click produces a list of reasons why that material would be defective, such as corrosion, scratch, and so forth. The inspector can select the defect type with the joystick or by speaking its name, and the information goes into the database. The user can navigate to the main selection screen by selecting or speaking on the "Main Menu" option on all screens. He can move one level up in the hierarchy with a single selection.

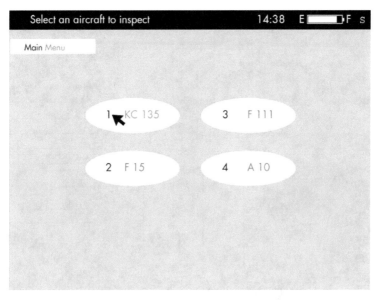

FIGURE 2.14: Navigator 2 aircraft selection.

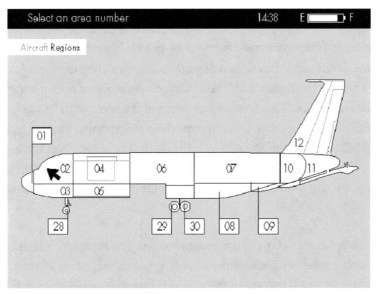

FIGURE 2.15: Navigator 2 region selection.

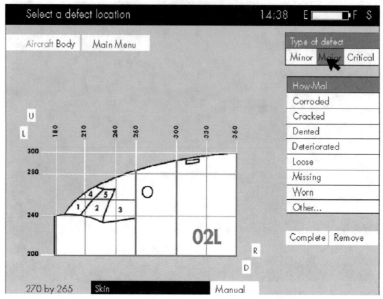

FIGURE 2.16: Sheet metal inspection screen.

The user-interface-design principles emphasized in design of the Navigator 2 include:

- Simplicity of function. The only functions available to the user are to identify skin imperfections for one of four aircraft types, transfer data to another computer, enter identification information for the vehicle and the inspector, and to see a screen that describes the Navigator 2 project.
- No text input. The user enters numbers to specify identification information. A special dialog was developed to enable entering numeric information using the joystick as an input device. This was cumbersome for users but was only required once per inspection.
- Controlled navigation. The interface was arranged as a hierarchy. The top level consisted of a menu with choice of function. Once the user chose an inspection function and then an aircraft, he navigated to the area of the skin by selecting an area of the aircraft to expand. When the user identified an imperfection, he had to select one of the allowable types of imperfections. At each stage, the user could go up one level of the hierarchy or return to the main menu.

Navigator 2: Lessons Learned. One of the lessons learned with Navigator 2 is the power of forcing the use of a common vocabulary. Because the average age of the aircraft is 35 years, the types of defects encountered are very well known. In the past, one inspector would call a defect "gouged," and another inspector would call the same defect a "scratch." What is the difference between a gouge and a scratch? How much material does a repair take? How much time? What skill is needed? The logistics of maintenance and repair are much more difficult without a standardized vocabulary. In this case, there was a serendipitous advantage to injecting more technology into the inspection process.

A second lesson is that in some cases, the speech-recognition interface mistakenly produced the wrong output. Speech-recognition systems typically have an error rate of 2% to 10%. This unexpected output may cause the application to produce the wrong result. In one of Navigator 2's early demonstrations, the user was attempting to exit the application. The speech-recognition system interpreted this input as a number. At that point the application was expecting a second number, but the user was saying "exit," "quit," " bye," and similar words. The system appeared to be frozen, but actually there was a mismatch between what the software was expecting and what the user thought the application's state was. In response to this problem, designers modified the interface to give the user more feedback on the state of the application. In addition, the team designed a novel application-input test-generator that took a description of the interface screens and created a list of all possible legal exits from each screen (Smailagic and Siewiorek, 1996).

A third lesson learned was the criticality of response time. When speech recognition was done in software on Navigator (ca. 1995), it took 12 times as long as real-time speech comprehension, and users became very frustrated by the wait. Moreover, people are less patient when they are on the move than when at a desktop. At a desk, a person is willing to wait 3 minutes for the computer to boot up, but when a person is on the move, his expectations are for instant response, like that of a portable tools such as a flashlight.

For example, some airplanes have a digital computer to control the passengers' overhead lights. The passenger is disconcerted when he presses the button and waits 2 or 3 seconds before the light turns on. Even a 2-second delay in a handheld device is disruptive. Users typically continue to push buttons until they get a response. These extra inputs cause a disconnect between the software and the user. The software receives a stream of inputs, but the user sees outputs that are related to inputs given a long time (e.g., several seconds) before the screen appears. The situation is similar to listening to yourself talk when there is a second or two delay in the sound played back. The user—or speaker—easily becomes confused.

Field evaluation of Navigator 2 indicated that the inspection is composed of three phases. The inspectors would spend the equivalent amounts of time (1) maneuvering their cherry picker to access a region of the airplane, (2) visually inspecting and feeling the airplane's skin, and (3) recording the defect's type and location. Navigator 2 reduced the paperwork time by half, resulting in an overall time savings of about 18% for the entire inspection. Training time to familiarize inspectors with the use of Navigator 2 was typically 5 minutes, after which they would proceed with actual inspections. (A major goal of field evaluations is that users perform productive work.)

Demomanship—what they do not teach you in school: The first public demonstration of Navigator 2 was at Boeing Aircraft in front of several VIPs. The hardware had been assembled in a marathon session. The software had been operational on a laptop computer for several weeks. Navigator 2 had a speech-activated interface. The system was operating smoothly until halfway through the demonstration when the system hung. It would not recognize any commands. We immediately thought it was a hardware problem. While one person talked about the design process and the upcoming field evaluation, the other took Navigator 2 apart looking for loose connectors. On reassembly, the system worked flawlessly.

Later we discovered the problem was caused by a design error. It took 3 days of intensive use before we could reproduce the problem. At only one point in the interface was a two-word sequence allowed (picking the region of the aircraft using a two-digit number, such as 02).

The user spoke a command to move to another screen, but the system recognized this as a number instead. While the system waited for a second number, the user spoke other commands in an attempt to exit the screen. There was a mismatch between the user's model of where the system was and where the software actually was. At this one place in all of the screens, there was no indication to the user of what to do next.

In effect, the speech-recognition system was acting as a random-input generator, making testing of the software very difficult. We later developed a way to take a description of the interface and automatically producing all legal inputs to each screen to determine if the system would hang on unexpected inputs. The test-generation system produced almost a million tests and found two defects that were not discovered in more than 6 months of system use. During the demonstration we could have saved a lot of time by simply rebooting the system. When in doubt, reboot!

A typical aircraft inspection requires about 36 hours and uncovers approximately 100 defects. Without Navigator 2, the inspector takes notes on a clipboard. When finished inspecting the aircraft, he fills out forms on a computer. Each defect takes 2 to 3 minutes to enter. The data entry is thus an additional 3- to 4-hour task. By contrast, Navigator 2 transmits the results of the inspection to a computer by radio in less than 2 minutes.

In summary, evaluations of inspectors before and after the introduction of Navigator 2 indicated a 50% reduction in the time to record inspection information (for an overall reduction of 18% in inspection time) and almost two orders of magnitude reduction in time to enter inspection information into the logistics computer (from more than 3 hours to 2 minutes). In addition, Navigator 2 weighs 2 pounds compared to the cart the inspectors currently use with 25 pounds of manuals.

Warning labels are no substitute for good design: As we raced to complete Navigator 2 for its field trials, we were unable to complete the circuit for the smart battery charger. The intention of the smart battery charger was to completely discharge the batteries before charging them again, thereby ensuring that the batteries were not overcharged. Because the circuit was not completely fabricated on time, we placed a warning label on the battery charger, stating that the charger should not be left on for more than 4 hours.

After the first day of field trials, the batteries were exhausted and were left in the charger overnight. The overcharged batteries exploded, spreading caustic gel around the inside of Navigator 2. Although the interior of the device was thoroughly cleaned, the acid from an exploded battery began eating away the metal connectors so that the on–off switch and other switches started to malfunction. Luckily, the corrosion process was slow enough that we were able to complete the 3-day field trial before the system became totally inoperable. The system's printed circuit board and internal metal stiffener had to be completely replaced.

2.4.5   Speech Translation

TIA-P System. The speech recognition/language translation (SR/LT) application consists of three phases: speech-to text-language recognition, text-to-text language translation, and text-to-speech synthesis. The application running on Tactical Information Assistant-Prototype (TIA-P; ca. 1996) is the Dragon Multilingual Interview System (MIS), jointly developed by Dragon Systems and the Naval Aerospace and Operational Medical Institute. It is a keyword-triggered multilingual playback system, which listens to a spoken phrase in English, proceeds through a speech recognition front-end, plays back the recognized phrase in English, and, after some delay (about 8 to 10 seconds), synthesizes the phrase in a foreign language (Croatian). The person listening to the translation can answer with "yes," "no," and some pointing gestures. The Dragon MIS software has about 45,000 active phrases, in the following domains: medical examination, minefields, road checkpoints, and interrogation. Therefore, key characteristics of this application are that it deals with a fixed set of phrases and it provides one-way communication. A similar system is used in Iraq as a briefing aid to interrogate former Iraqi intelligence officials and to speak with civilians about information relevant to locating individuals (Chisholm, 2004). This system shows the viability of the approach.

TIA-P is a commercially available system, developed by CMU, incorporating a 133-MHz 586 processor, 32 MB of DRAM, a 2-GB IDE hard disk, a full-duplex sound chip, and spread spectrum radio (2 Mbps, 2.4 GHz) in a ruggedized, handheld, pen-based system designed to support speech-translation applications (Figure 2.17).

FIGURE 2.17: TIA-P system.

The Dragon software loads into and stays in the system's memory. The translation uses un-compressed ~20 KB of .WAV files per phrase. There are two channels of output; the first plays in English, the second in Croatian. A stereo signal can be split, with one channel directed to an earphone and the second to a speaker. This is done in hardware attached to the external speaker. An Andrea noise-canceling microphone is used with an on–off switch.

Speech translation for one language (Croatian) requires a total of 60 MB of disk space. Speech recognition requires an additional 20–30 MB of disk space.

TIA-P has been tested with the Dragon speech-translation system in several foreign countries, including Bosnia (Figure 2.18), Korea, and Guantanamo Bay, Cuba. The system has also been used in human-intelligence data collection and experimentation with the use of electronic maintenance manuals.

TIA-P System: Lessons Learned. Field tests of the TIA-P system resulted in several valuable design lessons. The handheld display is quite convenient for checking the translated text. Use of the device revealed that wires should be kept to a minimum. Standard external electrical power should be available for use internationally, and the battery life should be extended. In addition, a design for ruggedness of the unit is important.

FIGURE 2.18: US soldier in the Balkans using TIA-P.

Example: Smart Modules. Another translation system is smart modules (ca. 1997), a family of wearable computers dedicated to speech processing (Smailagic et al., 2001). A smart module provides a service almost instantaneously and is configurable for different applications. The design goals for smart modules included reducing latency (translation time); removing context swaps (general-purpose computers time share between multiple applications, first loading an application, suspending execution and saving results to secondary storage, then loading the second application, and so forth); and minimizing weight, volume, and power consumption (Reilly, 1998).

The functional prototype of the smart module consists of two functionally specialized modules that perform language translation and speech recognition. The first module incorporates speech-to-text language recognition and text-to-speech synthesis. The second module performs text-to-text language translation. The LT module runs the PANLITE language-translation software (Frederking and Brown, 1996), and the SR module runs CMU's Sphinx II, which is a continuous, speaker-independent speech-recognition software (Ravishankar, 1996; Li et al., 1989), and Phonebox Speech Synthesis software.

Figure 2.19 shows the structure of the speech translator, from English to a foreign language, and vice versa. Speech input is done through the speech-recognition subsystem. A user wears a microphone as an input device, and background noise is eliminated with filtering procedures. A language model, generated from a variety of audio recordings and data, provides guidance for the speech-recognition system by acting as a knowledge source about the language properties.

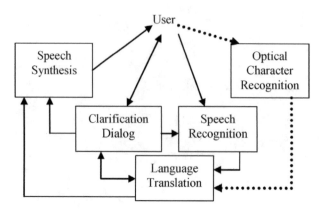

FIGURE 2.19: Speech translator system structure.

The language-translation engine uses an example-based machine translation (EBMT) system, which takes individual phrases and compares them to a set of examples in its memory to find phrases it can translate. A lexical MT (glossary) translates any unknown words that are left. The EBMT engine translates individual chunks of the sentence using the source-language model and then combines those fragments with a model of the target language to ensure correct syntax. When reading from the EBMT content, the system makes several random-access reads while searching for the appropriate phrase. Because random reads are done multiple times instead of loading large, continuous chunks of the content into memory, the disk latency times will be far more important than the disk bandwidth. The speech-generation subsystem performs text-to-speech conversion at the output stage. To make sure misrecognized words are corrected, a clarification dialog takes place onscreen; it includes the option to speak the word again or to write it in. As indicated in Figure 2.19, an alternative input modality could be the text from the optical character recognition subsystem (such as scanned documents in a foreign language), which is fed into the LT.

User-interface design for smart modules went through several iterations based on feedback from field tests. The emphasis was on getting completely correct two-way speech translation and having an easy-to-use, straightforward interface for the clarification dialog.

The team profiled and tuned the speech-recognition code. Profiling was performed to identify "hot spots" for hardware and software acceleration and to reduce the required computational and storage resources. A six-fold speedup was achieved over the original desktop computer system implementation of language translation, with five times smaller memory requirements (Christakos, 1998). Reducing operating system swapping and code optimization made a major impact in performance. Input to the smart module is audio and output is ASCII text. The speech-recognition module is augmented with speech synthesis. Figure 2.20 illustrates a combination of the language-

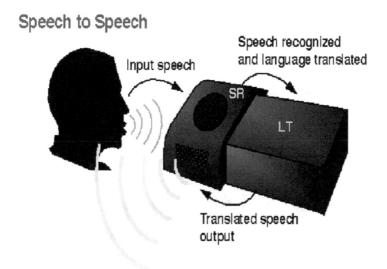

FIGURE 2.20: Speech recognizer (SR) and language translator (LT) smart module.

translation module (LT) and the speech-recognizer (SR) module, forming a standalone audio-based interactive dialog system for speech translation.

Target languages for the smart module included Serbo-Croatian, Korean, Creole French, and Arabic. Average language-translation performance was 1 second per sentence.

The key factors that determine how many processes can be run on a module are memory, storage space, and available CPU cycles. To minimize latency, the entirety of an application's working data set should be able to reside in memory.

Figure 2.21 shows the functional prototype of the Speech Translator Smart Module, with one module performing language translation and the other speech recognition and synthesis. In this prototype, the Apple Newton portable device completes the unit.

Figure 2.22 illustrates the response time for speech-recognition applications running on TIA-P and the Speech Recognition Smart Module. As the smart module is used with a lightweight operating system (Linux), compared to Windows 95 on TIA-P, and its speech-recognition code is more customized, the smart module has a shorter response time. An efficient mapping of the speech-recognition application onto the SR Smart Module architecture provided a response time very close to real time. It is important to provide feedback to the user in near real time to ensure system responsiveness.

SR Smart Module: Lessons Learned. The lessons learned from smart module tests and demonstrations include that a manual intervention process to correct misrecognized words resulted in some delays. In addition, swapping of data and programs can diminish the performance of the

FIGURE 2.21: Speech Translator Smart Module functional prototype.

language-translation module. Observation and feedback from users revealed that the size of the system's display can be as small as a deck of cards and still be effective.

The required system resources for speech-translator software in a smart module are several times smaller than comparable resources for a laptop or workstation version, as shown in Table 2.3.

2.4.6 Dual-Purpose Speech

In industry, most speech recognition on mobile computers concentrates on the tasks of form filling or simple interface commands and navigation. One reason these systems are not in wider use is that

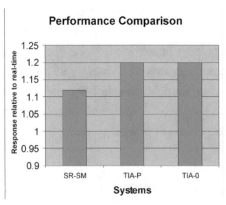

FIGURE 2.22: Response times for the smart module and TIA-P (lower is better).

TABLE 2.3: Comparison of required system resources			
	LAPTOP/ WORKSTATION	FUNCTIONAL MODULE SR/LT	OPTIMIZED MODULE SR/LT
Memory size	195 MB	53 MB	41 MB
Disk space	1 GB	350 MB	200 MB

speech interfaces are often socially interruptive when other people are nearby. Speech translation, such as the TIA-P system, is a different class of interface. In this type of application, the computer is an essential enabler of the conversation.

Lyons et al. (2004a) introduced a different type of conversation enabler in their dual-purpose speech work. Dual-purpose speech is easiest to discuss using a scenario. Tracy, a wearable computer user equipped with a head-mounted display and a Twiddler keyboard, is in conversation with a recently introduced colleague. Pressing a button on the keyboard, Tracy enables speech recognition and says "Bob, what is your phone number so that I have it for later?"

The wearable computer recognizes that its user wants to record a phone number and starts the user's contact list application. It attempts to recognize the name spoken and enters that into the application. The computer also captures the speech so that the user can correct the text later if there is an error.

Bob responds "Area code 404."

"404," repeats Tracy.

"555-1212," completes Bob.

"555-1212," continues Tracy, who presses another button on her keyboard to indicate that the interaction is over. "Ok, I have it!"

On Tracy's head-mounted display, a new contact has been created for Bob at (404) 555-1212. When Tracy finishes her conversation, she clicks an "accept" button on the application because she has recognized the information correctly. Tracy could also edit the information or play back the audio recorded during the interaction with Bob. Note that Tracy verbally repeated the information that Bob provided—a good conversational practice. Tracy both confirmed that she understood the information and provided Bob with an opportunity to correct her if necessary. This practice is also good from a privacy standpoint. Tracy wears a noise-canceling microphone, which is set to record only her voice and not that of her conversational partners. In this way, Tracy respects the privacy of her colleagues.

The Lyons team has designed dual-purpose speech applications for scheduling appointments, providing reminders for the user, and communicating important information to close colleagues. But the key point of this research for wearable computers is that these applications allow users to manipulate information on their wearables as part of the process of communicating (thus, the "dual-purpose" name). The users may actively format their speech so that the system can better understand them, and they may have to correct the system after an interaction. Yet the user manipulates the interface and enters information as part of a social process.

This style of interface provides a contrast to the traditional desktop computer, where the user's attention is assumed to be dedicated to the interface. Other wearable computing-related fields also attempt to create interfaces that are driven by the user's interactions with the environment. For example, the early augmented-reality systems by Feiner et al. (1993) attempted to display appropriate repair instructions based on the user's actions during the repair process. Such awareness of the user's context and goals may allow wearable computers to be used where a user's lack of attentional or physical resources would normally preclude traditional desktop applications.

2.5   PERCEPTION

Just as dexterity is impaired when a user is on the go, the user's ability to perceive a wearable computer's interface is lessened. The vibration and visual interference from a moving background interferes with visual tasks. Background noise and the noise from the body itself affect hearing. The moving of clothes over the body and the coupling of mechanical shock through the body can lessen the user's ability to perceive tactile displays. Sears et al. (2003) describe these detriments to mobile interaction caused by environmental and situational factors such as "situationally induced impairments and disabilities." These researchers and others are developing procedures to test human performance in mobile computing tasks in context (in this case, walking a path) (Barnard et al., 2005. Such research is sorely needed because not enough is known about how to adequately simulate mobile computing scenarios in testing. For example, in the work of Barnard et al. on performing reading comprehension tasks on personal digital assistants (PDAs) while walking, lighting levels affected workload measures more when the user was walking a path than when he was walking on a treadmill. The research community needs to achieve understanding about the interactions among mobility, attention, and perception in common mobile computing scenarios to adequately develop testing environments for mobile interfaces.

In the past, such work focused on cockpits, both for aviation and automobiles (Wickens, 2000; Melzer and Moffitt, 1997; Velger, 1998). However, the US military's Land Warrior project has highlighted the need for such research for soldiers who are walking (Blackwood, 1997).

Some researchers have begun exploring mobile output devices for very specific tasks. For example, Krum (2004) describes experiments with a head-up display that focus on determining how to render overhead views of an area to encourage learning the layout of the surrounding environment while the user is navigating to a goal on foot. As mobile-augmented reality is becoming practical from a technical standpoint, researchers have begun to address perceptual issues. Although not a mobile experiment, Laramee and Ware (2002) have investigated head-mounted displays to determine the relative effects of rivalry and visual interference between binocular and monocular displays with varying levels of transparency. As the market determines which mobile contexts are most important for users, experiments such as these will help determine how to design interfaces to least interfere with the user's primary tasks while providing the most value in terms of augmentation.

2.5.1   Example: eWatch, A Proactive Assistant

One promising direction for context-aware computing stems from recent research by the CMU team. This research uses unsupervised machine-learning techniques to combine real-time data from multiple sensors into a model of behavior that is individualized to the user. The eWatch is a wearable sensing, notification, and computing platform built into a wristwatch form factor. The wristwatch case make the device highly available, instantly viewable, ideally located for sensors, and unobtrusive to users (Smailagic et al., 2005). Bluetooth communication provides a wireless link to a cellular phone or stationary computer. eWatch senses light, motion, audio, and temperature and provides visual, audio, and tactile notification to the user. The system provides ample processing capabilities with multiple-day battery life, thereby enabling realistic user studies. Figure 2.23 shows representative eWatch screenshots—sensor waveforms, calendar, and messages—and the watch itself.

Figure 2.24 illustrates sensors and main hardware components of eWatch.

Knowing about the user's location is an important aspect of a context-aware system. Using eWatch, we developed a system that identifies previously visited locations. Our method uses information from the audio and light sensor to learn and distinguish different environments. We recorded and analyzed the audio environment and the light conditions at several different locations. Experiments showed that locations have unique background noises, such as traffic, talking, and the noises of computers, air-conditioning, and television.

The light sensor sampling at a high frequency can provide additional information beyond the brightness of the location. Frequency components generated by fluorescent lights (at 120 and 240 Hz) and displays (television, computer screen at 60 Hz) can be observed, as shown in Figure 2.25. (Incandescent bulbs do not have a frequency component.)

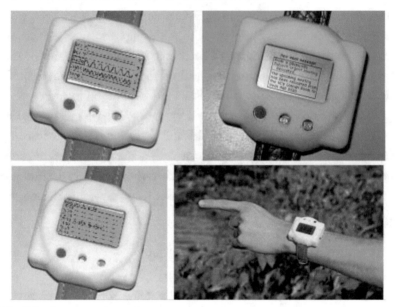

FIGURE 2.23: eWatch, a proactive assistant.

We observed that the frequency characteristics of light conditions tend to remain constant in most locations. For our study, audio data were recorded with the built-in microphone at a sample rate of 8 kHz and the light sensor at a frequency of 2048 Hz. Locations frequently visited by the user were recorded; these locations included the rooms of the user's apartment (living room, kitchen, bedroom, bathroom), office, computer laboratory, different street locations on the way to university,

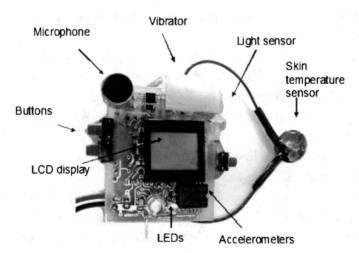

FIGURE 2.24: eWatch internal hardware.

the interior of a bus, several restaurants, and a supermarket. Each location was visited multiple times on different days. The classification with the light sensor alone achieved 84.9% correctly classified samples. The overall recognition accuracy was 87.4% when using only the audio sensor. Both sensors combined gave the best result of 91.4% (Maurer et al., 2006a).

We also used the eWatch accelerometer to identify physical activity, focusing on classification of six primary activities: sitting, standing, walking, ascending stairs, descending stairs, and running. Our experiments have demonstrated the sensitivity of various body positions for detecting a range of user activities. We tested the eWatch in several body positions that are normally used for wearing electronic devices such as cell phones or PDAs. Our eWatch sensor hardware was placed on the left wrist, belt, a necklace, in the right trouser pocket, shirt pocket, and bag. The subjects wore six eWatch devices located at these body positions during the study. The devices recorded sensor data from the accelerometer and light sensor into their flash memory. The user was asked to perform tasks such as working on the computer or walking to another building. eWatch recorded both axes of the accelerometer and the light sensor. We calculated features from both accelerometer axes (X and Y), the light sensor, and a combined value of both accelerometer signals. The decision-tree classifier was chosen and used because it provides a good balance between accuracy and computational complexity.

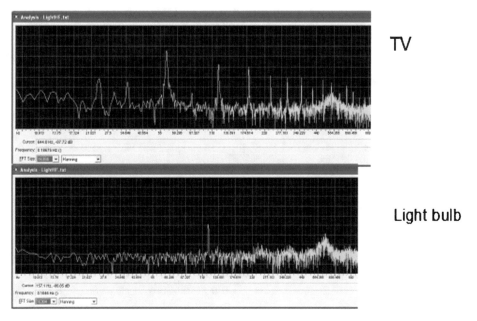

TV

Light bulb

FIGURE 2.25: Sensing context, light sensor.

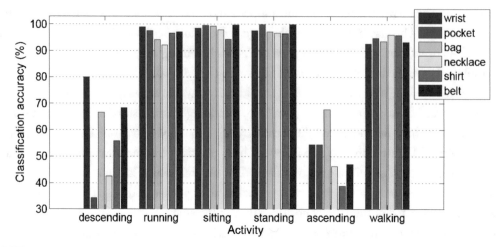

FIGURE 2.26: Recognition accuracy for typical activities at several different body locations.

We calculated the classification accuracy for every activity in the six body positions. The data from all subjects were combined for the training of a general classifier that is not specific to a person.

Figure 2.26 shows the recognition accuracy for the individual activities at different body locations. The data indicate that any of the six positions is good for detecting *walking*, *standing*, *sitting*, and *running*. *Ascending* and *descending* the stairs is difficult to distinguish from *walking* in all positions because the classifier was trained for multiple persons. We found that the wrist performs best for descending, and the bag best for ascending.

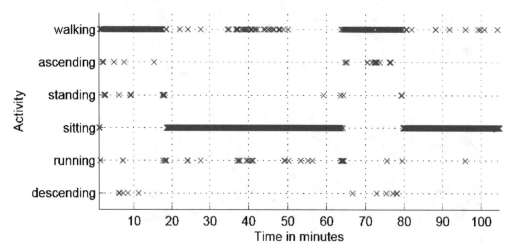

FIGURE 2.27: Activity classification recorded over 100 minutes.

On the basis of these results, we implemented a decision-tree classifier that runs on the eWatch. The classification results are stored into flash memory and are downloaded to a computer later for further processing and analysis. The feature extraction and decision-tree classification are implemented on eWatch.

The figure above shows classification of activities for one person. A subject wore the eWatch with the built-in activity classifier on the wrist. The system classified the activity in real time and recorded the classification results to flash memory. Figure 2.27 shows 100 minutes of activity classification, as the user walked to a restaurant, sat down, ate lunch, went back to the office, and sat down to continue working. The classification results match well with the actual activities; eating lunch was partially interpreted as walking or running activity because of the subject's arm movements.

CHAPTER 3

Design Guidelines for Wearable Computing

The combined studies and research reported in this lecture suggest a number of useful guidelines for designing wearable computing devices. These guidelines are summarized below. Also included with the guidelines is a list of questions that designers should consider when beginning to design a wearable computer.

Device design:

- Keep the user at the center of the design.
- Use an iterative design process that includes developing a prototype, having users evaluate it, and incorporating user feedback into a revised design.
- Always design for the comfort of the user.
- Design for a human scale; for example, consider users' finger sizes for buttons or keypads and hand sizes for palm-held devices.
- Similarly, design for the variety of sizes that users are.
- For devices that will be passed from one user to another, such as a wrist-mounted data-input device, design detachable components, such as a wrist strap, so that each user has the appropriate size and so that these parts of the device can be cleaned easily.
- Make controls obvious and design them so that they are usable from whatever position on the body a user wears the device.
- Be aware of which areas of the body work well for wearable devices and which do not.
- Keep wearable devices to a personal scale—not too large or too heavy for comfortable prolonged use.
- Make controls easy to read, with the same feel and orientation regardless of the device's position on the user's body.

Interface design:

- Design the interface so that navigation is simple and clear.

- Make controls easy to read, with the same feel and orientation regardless of the device's position on the user's body.
- Design the interface so that text input is not needed; use buttons, dials, or voice for the user to select options in the device's interface.
- Limit the navigation's directions and options; use a hierarchy with clear paths between levels and a persistent link to the main menu or screen.

Designing for the user's environment:

- Design components that have sufficient ruggedness for the conditions in which they will be used.
- Plan the device so that access to controls and viewing of the interface are easily maintained even when the user—and the device—are moving.
- Similarly, design the controls and interface so that the user can readily see them even when his or her attention must be divided between the exterior environment and the wearable computing task.
- Design to accommodate the cultural and esthetic standards of the community in which the device will be used.
- Design the device so that it can easily make a connection with an external system to download data entered by the user.

Questions designers should ask when beginning a project:

- Does the mental model match the application physical workflow?
- Are the user interactions with the application simple and intuitive?
- Are the input and output devices and modalities appropriate for the set of projected applications?
- Does the system have enough resources (processor, memory, network) to be responsive to the user's interactions without excess?
- Will the form and shape of the wearable computer be comfortable with the movement of the human body?
- Where will the computer be placed on the body? What is the size, weight, and power consumption?
- Is the device's thermal management and heat dissipation appropriate for the intended physical environment?

CHAPTER 4

Research Directions

The evolution of computing has shown that it takes several years to develop a user interface style and that this often emerges quite a while after the technology threshold has been passed. The thresholds represent the time when microprocessors have the capability of supporting the indicated form of interface. Figure 4.1 depicts the increase in microprocessor performance (measured in millions of instructions per second, or MIPS) as a function of time. In the early 1960s, Gordon Moore of Intel made the observation-prediction that the capacity of semiconductor chips was doubling every 2 years (Moore's law). Similar trends have been noted for microprocessor speed, disk storage capacity, and network bandwidth. The points depicted in Figure 4.1 are the performance thresholds necessary for each type of user interface. Thus, a textual interface requires 1 MIPS, a graphical user interface, 10 MIPS; a handwriting interface, 30 MIPS; a speech recognition interface, 100 MIPS; a natural language understanding, 1000 MIPS; and vision understanding, 10,000 MIPS of computing power.

Because ease of use is so closely associated with human reaction, it is much more difficult to quantify. There are at least three basic functions related to ease of use, however: input, output, and information representation. Figure 4.2 summarizes several points for each of these basic functions.

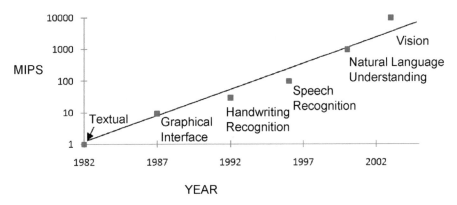

FIGURE 4.1: User interface performance thresholds.

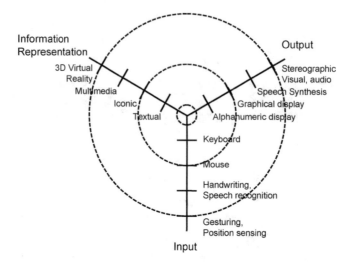

FIGURE 4.2: Kiviat graphs for wearable computer use modalities.

Note that unlike the continuous variables for capacity and performance, the ease-of-use metrics are discrete.

Just as the performance of microprocessors has increased over time (as shown in Figure 4.1), the characteristics of the user interface shown in Figure 4.2 are also moving out with time. For example, the keyboard with an alphanumeric display using textual information is representative of time-sharing systems of the early 1970s. The keyboard and mouse, graphical output, and iconic desktop are representative of personal computers of the early 1980s. The addition of handwriting-recognition input, speech-synthesis output, and multimedia information began emerging in the early 1990s. It takes approximately one decade to broadly disseminate new input, output, and informational representations. In the 2000s, speech recognition, position sensing, and eye tracking are becoming common inputs. Heads-up projection displays should allow superposition of information onto the user's environment.

4.1 WEARABLE COGNITIVE AUGMENTATION

Our major goal of research is to determine the cognitive state of a user—especially the user's cognitive load—from external observations. Knowing the user's cognitive state would enable development of proactive cognitive assistants that anticipate user needs much like a human assistant does.

What makes this attempt possible is an unprecedented advance in measuring and understanding brain activity during complex tasks using functional magnetic resonance imaging (fMRI), as depicted in Figure 4.3. The brain activity measured with fMRI is only one step removed from the neural activity itself. fMRI provides a measure of the oxygenated hemoglobin in the capillary

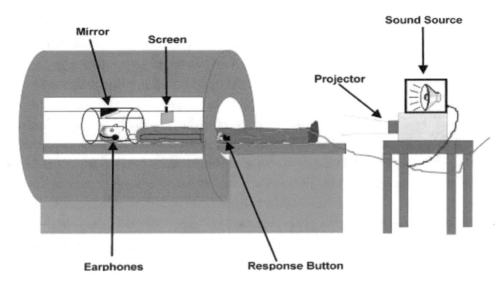

FIGURE 4.3: fMRI experiment configuration.

beds in which the neural activity is occurring. Routinely used protocols in neurocognitive research on advanced MRI scanners sample the entire cortex approximately once per second. It is feasible to pursue a research plan to recognize some of the brain-cognitive states that should be amenable to improvement, and then to develop an intelligent tutoring system that uses the fMRI-measured

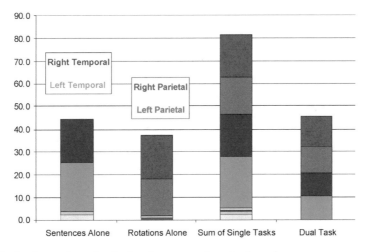

FIGURE 4.4: Activation volume in dual task compared to single task.

brain activation to guide the tutoring, to infer current mental states and to rapidly guide the learner to desired mental states.

There is also a maximum on the total activation across cortical areas. Such a system-wide capacity constraint might be expected to operate when subjects perform two tasks that draw on non-overlapping brain areas. The requirement of nonoverlap ensures that any constraint on performance is not just a result of competition for the same neural mechanisms. In a study that found evidence for such a constraint, the two tasks were auditory sentence comprehension and mental rotation (Just et al., 2001). If there were no system-wide capacity constraint, one would expect that because the two tasks draw on different neural substrates (language- and spatial-related areas, respectively), the activation in the dual task would simply be the union of the activations in each of the two single tasks. However, the activation in the dual task was far less than the union of the two single tasks. The activation associated with each individual task decreased by 30% to 50% in the dual-task condition and in the graph showing the group data results (Figure 4.4). The decrease in the dual-task condition applied to 17 of the 18 subjects. Thus, there appears to be a detectable upper boundary on the total amount of activation that can be sustained in a set of cortical areas.

CHAPTER 5

Conclusions and Future Challenges

Wearable computers are an attractive way to deliver a ubiquitous computing system's interface to a user, especially in non-office-building environments. The biggest challenges in this area deal with fitting the computer to the human in terms of interface, cognitive model, contextual awareness, and adaptation to tasks being performed. These challenges include:

- User interface models. What is the appropriate set of metaphors for providing mobile access to information (i.e., what is the next "desktop" or "spreadsheet")? These metaphors typically take a decade or longer to develop (the desktop metaphor started in early 1970s at Xerox PARC and required more than a decade before it was widely available to consumers). Extensive experimentation working with end-user applications will be required. Furthermore, there may be a set of metaphors each tailored to a specific application or a specific information type.

- Input/output modalities. Although several modalities mimicking the input/output capabilities of the human brain have been the subject of computer science research for decades, the accuracy and ease of use (many current modalities require extensive training periods) are not yet acceptable. Inaccuracies produce user frustrations. In addition, most of these modalities require extensive computing resources, which will not be available in low-weight, low-energy wearable computers. There is room for new, easy-to-use input devices such as the dial developed at CMU for list-oriented applications.

- Quick interface evaluation methodology. Current approaches to evaluate a human–computer interface require elaborate procedures, with scores of subjects. Such an evaluation may take months and is not appropriate for use during interface design. These evaluation techniques should especially focus on decreasing human errors and frustration.

- Matched capability with applications. The current thought is that technology should provide the highest performance capability. However, this capability is often unnecessary to complete an application, and enhancements such as full-color graphics require substantial resources and may actually decrease ease of use by generating information overload for the user. Interface designers and evaluators should focus on the most effective means for information access and resist the temptation to provide extra capabilities simply because they are available.

- Context-aware applications. Among the questions to be addressed as we develop context-aware applications are as follows: How do we develop social and cognitive models of applications? How do we integrate input from multiple sensors and map them into user social and cognitive states? How do we anticipate user needs? How do we interact with the user? Some initial results have been reported in by Krause et al. (2006).

- Proactive assistant. As designs for a proactive assistant are developed, we must address questions such as these: When and how to interrupt a user? What information is collected and how is privacy protected? Who can access the information? How long is information stored? How do users specify preferences for data availability? And finally, can we charge market value to those demanding attention?

References

Anhalt, J., Smailagic, A., Siewiorek, D., et al. (2001). Towards context aware computing. *IEEE Intelligent Systems, 6*(3), 38–46. doi:10.1109/5254.940025

Azuma, R. (1997). A survey of augmented reality. *Presence, 6*(4), 355–386.

Barnard, L., Yi, J. S., Jacko, J. A., and Sears, A. (2005). An empirical comparison of use-in-motion evaluation scenarios for mobile computing devices. *International Journal of Human Computer Studies, 62*, 487–520. doi:10.1016/j.ijhcs.2004.12.002

Barnard, L., Yi, J., Jacko, J. A., and Sears, A. (2007). Capturing the effects of context on human performance in mobile computing systems. *Personal and Ubiquitous Computing, 11*(2), 81–96. doi:10.1007/s00779-006-0063-x

Blackwood, W. (1997). *Tactical display for soldiers*. National Academy of Sciences, Washington, DC.

Blaskó, G., Feiner, S., and Coriand, F. (2005). Exploring interaction with a simulated wrist-worn projection display. *Proceedings of the Ninth IEEE International Symposium on Wearable Computers*. IEEE Computer Society Press, Los Alamitos, CA, pp. 2–9.

Bodine, K., Gemperle, F. (2003). Effects of functionality on perceived comfort of wearables. *Proceedings of the Seventh IEEE International Symposium on Wearable Computers*. IEEE Computer Society Press, Los Alamitos, CA, pp. 57–61. doi:10.1109/ISWC.2003.1241394

Chamberlain, A., Kalawsky, K. (2004). A comparative investigation into two pointing systems for use with wearable computers while mobile. *Proceedings of the Eighth IEEE International Symposium on Wearable Computers*. IEEE Computer Society Press, Los Alamitos, CA, pp. 110–117. doi:10.1109/ISWC.2004.1

Chisholm, M. (2004). Technology that speaks in tongues. *Military Information Technology, 8*(2). Retrieved from http://www.military-information-technology.com/article.cfm?DocID=424.

Christakos, C.K. (1998). *Optimizing a language translation application for mobile use*. Master's thesis, CMU, Department of Electrical and Computer Engineering, Pittsburgh, PA.

Clawson, J., Lyons, K., Starner, T., and Clarkson, E. (2005). The impacts of limited visual feedback on mobile text entry using the mini-QWERTY and Twiddler keyboards. *Proceedings of the IEEE International Symposium on Wearable Computers*. IEEE Computer Society Press, Los Alamitos, pp. 170–177. doi:10.1109/ISWC.2005.49

Feiner, S., MacIntyre, B., and Seligmann, D. (1993). Knowledge-based augmented reality. *Communications of the ACM, 36*(7), 52–62. doi:10.1145/159544.159587

Frederking, R. E., Brown, R. (1996). The Pangloss-lite machine translation system: expanding MT horizons. *Proceedings of the Second Conference of the Association for Machine Translation in the Americas.* Montreal, Canada, pp. 268–272.

Gemperle, F., Kasabach, C., Stivoric, J, Bauer, B., and Martin, R. (1998). Design for wearability. *Second International Symposium on Wearable Computers.* IEEE Computer Society Press, Los Alamitos, CA, pp. 116–122.

Hong, J.I., Landay, J.A. (2004). An architecture for privacy-sensitive ubiquitous computing. *Proceedings of the Second International Conference on Mobile Systems, Applications, and Services.* IEEE Computer Society Press, Los Alamitos, CA, pp. 177–189. doi:10.1145/990064.990087

Iachello, G., Smith, I., Consolvo, S., Abowd, G., et al. (2005). Control, deception, and communication: evaluating the deployment of a location-enhanced messaging service. *Proceedings of the Seventh International Conference on Ubiquitous Computing.* IEEE Computer Society Press, Los Alamitos, CA, pp. 213–231.

Just, M. A., Carpenter, P. A., Keller, T. A., Emery, L., Zajac, H., and Thulborn, K. (2001) Interdependence of nonoverlapping cortical systems in dual cognitive tasks. *NeuroImage, 14,* 417–426. doi:10.1006/nimg.2001.0826

Kollmorgen, G. S., Schmorrow. D., Kruse, A., and Patrey, J. (2005). The cognitive cockpit state of the art human–system integration. *Proceedings of the 2005 Interservice/Interindustry Training Simulation and Education Conference.* DARPA, Arlington, VA, pp. 57–61.

Krause, A., Smailagic, A., and Siewiorek, D.P. (2006). Context-aware mobile computing: learning context-dependent personal preferences from a wearable sensor array. *IEEE Transactions on Mobile Computing, 5*(2), 113–127.

Krum, D., Wearable computers and spatial cognition, Ph.D. Thesis, Georgia Institute of Technology, Atlanta, GA, 2004.

Li, K. F., Hon, H.W., Hwang, M.J., and Reddy, R. (1989). The Sphinx speech recognition system. *Proceedings of the IEEE ICASSP.* IEEE Computer Society Press, Los Alamitos, pp. 170–177.

Laramee, R., Ware, C. (2002). Rivalry and interference with a head mounted display. *ACM Transactions on Computer–Human Interface, 9*(3), 238–251. doi:10.1145/568513.568516

Lyons, K.., Starner, T. (2001). Mobile capture for wearable computer usability testing. *Proceedings of the International Symposium on Wearable Computers.* IEEE Computer Society Press, Los Alamitos, pp. 170–177. doi:10.1109/ISWC.2001.962099

Lyons, K, ,Skeels, C., Starner T., Snoeck, B., Wong, B., and Ashbrook, D. (2004a). Augmenting conversations using dual-purpose speech. *Proceedings User Interface and Software Technology.* Montreal, Canada, pp. 237–246. doi:10.1145/1029632.1029674

Lyons K., Starner, T., and Plaisted, D. (2004b). Expert typing using the Twiddler one handed chord keyboard. *Proceedings of the IEEE International Symposium on Wearable Computers.* Montreal, Canada, pp. 94–101.

Martin, T. (1999). *Balancing batteries, power, and performance: system issues in CPU speed-setting for mobile computing.* PhD thesis, Carnegie Mellon University, Department of Electrical and Computer Engineering, Pittsburgh, PA.

Maurer, U., Rowe, A., Smailagic, A., and Siewiorek, D. (2006a). eWatch: a wearable sensor and notification platform. *IEEE International Workshop on Wearable and Implantable Body Sensor Networks.* IEEE Computer Society Press, Boston, MA, pp. 113–116. doi:10.1109/BSN.2006.24

Maurer, U., Smailagic, A., and Siewiorek, D. (2006b). Activity recognition and monitoring using multiple sensors on different body positions. *IEEE International Workshop on Wearable and Implantable Body Sensor Networks.* IEEE Computer Society Press, Boston, MA, pp. 142–145. doi:10.1109/BSN.2006.6

Melzer, J., Moffitt, K. (1997). *Head mounted displays: designing for the user.* McGraw-Hill, New York.

Mizell, D. (2001). *Boeing's wire bundle assembly project: fundamentals of wearable computers and augmented reality.* Barfield , W., Caudell, T., Eds. Lawrence Eribaum & Associates, NJ, pp. 447–467.

Ockerman, J. (2000). *Task guidance and procedure context: aiding workers in appropriate procedure following.* Ph D thesis, Georgia Institute of Technology, School of Industrial System Engineering, Atlanta, GA.

Oulasvirta, A., Tamminen, S., Roto, V., and Kuorelahit, J. (2005). Interaction in 4-second bursts: the fragmented nature of attentional resources in mobile HCI. *Proceedings of the SIGCHI Conference on Human Factors in Computing Systems.* Montreal, Canada, pp. 919–928. doi:10.1145/1054972.1055101

Ravishankar, M. (1996). *Efficient algorithms for speech recognition.* PhD thesis, Carnegie Mellon University, School of Computer Science, Pittsburgh, PA.

Reilly, D. (1998). *Power consumption and performance of a wearable computing system.* Master's thesis, Carnegie Mellon University, Electrical and Computer Engineering Department, Pittsburgh, PA.

Sears, A., Lin, M., Jacko, J., and Xiao, Y. (2003). When computers fade: pervasive computing and situationally-induced impairments and disabilities. *Proceedings of HCII 2003.* Montreal, Canada, pp. 1298–1302.

Siewiorek, D.P. (2002). Issues and challenges in ubiquitous computing: new frontiers of application design. *Communications of the ACM, 45*(12), 79–82.

Siewiorek, D.P., Smailagic, A., and Lee, J. C. (1994). An interdisciplinary concurrent design methodology as applied to the Navigator wearable computer system. *Journal of Computer and Software Engineering, 2*(2), 259–292.

Siewiorek, D.P, Smailagic, A., Bass, L., Siegel, J., Martin, R., and Bennington, B. (1998). Adtranz: a mobile computing system for maintenance and collaboration. *Proceedings of the Second IEEE*

International Conference on Wearable Computers. Montreal, Canada, pp. 25–32. doi:10.1109/ISWC.1998.729526

Smailagic, A. (1997). ISAAC: a voice activated speech response system for wearable computers. *Proceedings of the IEEE International Conference on Wearable Computers.* Montreal, Canada, pp. 183–184. doi:10.1109/ISWC.1997.629945

Smailagic, A., Siewiorek, D. (1993). A case study in embedded system design: the VuMan 2 wearable computer. *IEEE Design and Test of Computers, 10*(3), 56–67. doi:10.1109/54.232473

Smailagic, A., Siewiorek, D. P. (1994). The CMU mobile computers: a new generation of computer systems. *Proceedings of the IEEE COMPCON, 94.* IEEE Computer Society Press, Los Alamitos, CA, pp. 467–473. doi:10.1109/98.486972

Smailagic, A., Siewiorek, D. (1996). Modalities of interaction with CMU wearable computers. *IEEE Personal Communications, 3*(1), 14–25.

Smailagic A., Siewiorek D.P., Martin R., and Stivoric, J. (1998). Very rapid prototyping of wearable computers: a case study of VuMan 3 custom versus off-the-shelf design methodologies. *Journal of Design Automation for Embedded Systems, 3*(2–3), 219–232.

Smailagic, A., Siewiorek, D. P., and Reilly, D. (2001). CMU wearable computers for realtime speech translation, *IEEE Personal Communications, 8*(2), 6–12.

Smailagic, A., Siewiorek, D.P., Maurer, U., Rowe, A., and Tang, K. (2005). eWatch: context sensitive system design case study. *Proceedings of the Annual IEEE VLSI Symposium.* IEEE Computer Society Press, Los Alamitos, CA, pp. 98–103. doi:10.1109/ISVLSI.2005.31

Smailagic, A., Siewiorek, D.P. (2007). MoRE : Mobile Reference Environment, Technical Report Institute for Complex Engineered Systems, Carnegie Mellon University, Pittsburgh, PA, 2007.

Starner, T., Snoeck, C., Wong, B., and McGuire, R., (2004). Use of mobile appointment scheduling devices. *Proceeding of the ACM Conference Human Factors in Computing Systems.* Montreal, Canada, pp. 1501–1504. doi:10.1145/985921.986100

Starner, T. (2001). The challenges of wearable computing: part 1+2, *IEEE Micro, 21*(4), 44–67.

Stein, R., Ferrero, S., Hetfield M., Quinn, A., and Krichever, M. (1998). Development of a commercially successful wearable data collection system, *Proceedings of the IEEE International Symposium on Wearable Computers.* Montreal, Canada, pp. 18–24.

Velger, M. (1998). *Helmet-mounted displays and sights.* Artech House, Inc., Norwood, MA.

Wickens, C. D. (2000). Imperfect and unreliable automation and its implication for attention allocation, information access and situation awareness, Final Technical Report (ARL-00-10 / NASA-00-2), Aviation Research Lab., University of Illinois, Savoy.

Author Biography

Daniel P. Siewiorek received his bachelor of science degree in electrical engineering from the University of Michigan, Ann Arbor, in 1968 and his master of science and doctor of philosophy degrees in electrical engineering, with a minor in computer science, from Stanford University in 1969 and 1972, respectively.

Dr. Siewiorek is the Buhl University professor of electrical and computer engineering and computer science at Carnegie Mellon University. He has designed and been involved with the design of nine multiprocessor systems and has been a key contributor to the dependability design of more than two dozen commercial computing systems. Dr. Siewiorek leads an interdisciplinary team that has designed and constructed more than 20 generations of mobile computing systems. He has written eight textbooks in the areas of parallel processing, computer architecture, reliable computing, and design automation in addition to more than 475 papers. Dr. Siewiorek has served as associate editor of the Computer System Department of the Communications of the Association for Computing Machinery, as chairman of the IEEE Technical Committee on Fault-Tolerant Computing, and as founding chairman of the IEEE Technical Committee on Wearable Information Systems. Currently director of the Human Computer Interaction Institute, he was previously director of the Engineering Design Research Center and cofounder of its successor organization, the Institute for Complex Engineered Systems, where he served as associate director. He has been the recipient of the American Association of Engineering Education Frederick Emmons Terman Award, the IEEE/ACM Eckert–Mauchly Award, and the ACM SIGMOBILE Outstanding Contributions Award. He is a fellow of IEEE, ACM, and AAAS and a member of the National Academy of Engineering.

Asim Smailagic is a research professor in the Institute for Complex Engineered Systems, College of Engineering, and Department of Electrical and Computer Engineering at Carnegie Mellon University (CMU). He is also the director of the Laboratory for Interactive and Wearable Computer Systems at CMU, which has designed and built more than two dozen generations of novel wearable computers during the last 15 years and several prototypes of context-aware computer systems. Dr. Smailagic received the Fulbright Postdoctoral Award in Computer Science at CMU in 1988. He has been a program chairman or cochairman of IEEE conferences more than 10 times. He is chair of the IEEE Technical Committee on Wearable Information Systems. Dr. Smailagic has acted as coeditor, associate editor, and guest editor in leading archival technical journals, such as *IEEE Transactions on Mobile Computing, IEEE Transactions on VLSI Systems, IEEE Transactions on Parallel and Distributed Systems, IEEE Transactions on Computers, EURASIP Journal on Embedded Systems, Pervasive and Mobile Computing, Journal on VLSI Signal Processing*, and others. He codeveloped an interdisciplinary concurrent design methodology with Dr. Dan Siewiorek and is widely recognized for his work in the design and rapid prototyping of wearable computers. Dr. Smailagic received the 2000 Allen Newell Award for Research Excellence from CMU School of Computer Science, the 2003 Carnegie Science Center Award for Excellence in Information Technology, the 2003 Steve Fenves Systems Research Award from the CMU College of Engineering, and other awards. Dr. Smailagic has written or edited books in the areas of mobile computing, digital system design, field programmable gate arrays, and VLSI systems. He gave keynote lectures at many representative international conferences and at institutions such as the Royal Academy of Engineering, London, UK. Prof. Smailagic participated in several major research projects that represent milestones in the evolution of computer system architectures: from CMU's Cm* Multiprocessor System and Edinburgh Multi-Microprocessor Assembly (EMMA) to CMU's parallel and distributed computer systems, to the current projects on Wearable Computer Systems, Smart Modules computers, Aura Pervasive Computing, Context Aware Computing, Cognitive Computing, Virtual Coaches, and Quality of Life Technologies.

Thad Starner is the founder and director of the Contextual Computing Group at Georgia Institute of Technology's College of Computing. Before joining the Georgia Tech faculty in 1999, Starner gained international recognition at the MIT Media Laboratory as one of the world's leading experts on wearable computers during his doctoral work "Wearable Computing and Contextual Awareness." An advocate of continuous-access, everyday-use systems, Starner has worn his custom wearable computer in such a manner since 1993, arguably the longest such experience. Starner is a cofounder of the IEEE International Symposium on Wearable Computers (ISWC) and cofounder and first member of the MIT Wearable Computing Project. In 1999, Starner was named one of Technology Review's TR100—100 individuals under 35 who exemplify the spirit of innovation. Starner has been a keynote speaker and/or distinguished lecturer at a wide range of scientific, industrial, and academic events, including ACM's Conference for Advances in Computer Entertainment (ACE), Princeton's Distinguished Lecture in Pervasive Computing, the International Conference on Robotics and Automation (ICRA), Fashion Institute of Technology's Faculty Convocation, Interaction, Nicograph, Second Congreso Internacional de Ingenieria Electronica, and Interaction Homme–Machine (IHM). His work has been discussed in national and international public forums, including CBS's *60 Minutes* and *48 Hours*, *The New York Times*, *New Scientist*, *Nikkei Science*, ABC's *Nightline* and *World News Tonight with Peter Jennings*, *The London Independent*, *The Bangkok Post*, PBS's *Scientific American Frontiers*, CNN, BBC, and *The Wall Street Journal*. Starner has authored well over 100 scientific articles and book chapters and is always looking for a good game of table tennis.

Printed in the United States
by Baker & Taylor Publisher Services